Annihilation...

Annihilation...
The Universal Theorem

David Tovell

Published in Canada by Panache,
Toronto, Ontario.

Annihilation... The Universal Theorem (TUT)
ISBN 978-0-9865702-2-3 (paperback)
ISBN 978-0-9865702-1-6 (e-book)

Cover Design: Cindy Cake, WeMakeBooks.ca
Text Design & Layout: Beth Crane, WeMakeBooks.ca
Figure Design: Stephen Waicus
Printed by CreateSpace

Cover: This spiral galaxy selected is in the constellation Cepheus. The distance to M100 from earth has been calculated to be approximately 56 million light years, and the galaxy has a system of gas and stars similar to our own Milky Way.

Available from Amazon.com, CreateSpace.com, and other retail outlets
Available also on Kindle and other devices

In Dedication to
Lawrence B. (Larry) Sawyer
1911–2002

This essay is dedicated to my mentor, Lawrence Sawyer, who told his body of students many years ago, 'My daddy told me when I was a young boy that when I grew up to be man, I would either be very rich or be a philosopher. Well, I'm a philosopher!'.

Over my auditing career, I met with Larry frequently and he always added some wonderful philosophical insights and inspiration to my life and career. He also had a great influence on his students that included my auditing interns. His writings and other works ultimately led to the establishment of our global auditing profession of which I am proud to be a member.

If I neglected to thank Larry appropriately during those many years, then, according to 'The Universal Theorem', I will still have sufficient opportunities to do so in the future universes, which we will be sharing, and probably auditing!

David Tovell, CGA, CIA
Toronto, Ontario
2013

Contents

Author's Note

I wrote *Annihilation...The Universal Theorem* to answer the universal and philosophical question, 'Why'. After reading the book *The Grand Design* by Stephen Hawking and Leonard Mlodinow, I was captivated by their natural question 'Why':

> Luck in the precise form and nature of fundamental physical law is a different kind of luck from luck we find in environmental factors. It cannot be so easily explained, and has far deeper physical and philosophical implications. Our universe and its laws appear to have a design that both is tailor-made to support us and, if we are to exist, leaves little room for alteration. That is not easily explained, and raises the natural question of why it is that way.

My belief is that The Universal Theorem (TUT) explains the philosophical implications of the statement, '... *raises the natural question of why it is that way*'. Our collective existence in our universe is based on annihilation and reconstruction of universes over infinity. The perpetual frequency of these events ensures a short duration of universes supporting our existence. My philosophical conclusion based on TUT is whatever was, is!

When reading the essay, I would suggest not skipping the terminology section, and referring back to it as needed. Since each section is built upon those that came before, it would be wise to not skip segments, but read chronologically.

Terminology

It is important that The Universal Theorem (TUT) and its related philosophies, described in the Introduction and The Philosophical Context, be clearly communicated. The language used is common but when thinking and writing outside the box, the meanings have a different context. The purpose of the terminology is connecting the common words used in a different context for application in TUT and elaborating on their intended meaning.

Annihilation: Where all stable spiral path segments in the universe transfer in a cascade, with all of their respective cohorts of existence on parallel time lines intersects, to points of non-existence.

Axis of direction: Axis is generally understood as a rotation around a straight line but in TUT only time is defined as a straight line. Therefore, the axis is a curved line. The direction refers to the axis of rotation on a curved line segment that leads to infinity.

Cohorts: Points of existence sharing parallel time lines.

Corporate body: The body of unified individual biological cells.

Doppler effect: According to TUT, the normal energy spectrum is where the object and observer are on stable spiral path segments and where the energy spectrum transfers are constant. All changes in the energy spectrum are due to the variability of spiral path segments. The Doppler effect, in universal terms, is a contrast between the energy spectrum in stable spiral path segments and the energy spectrum in variable spiral path segments.

Energy: Any change in a spiral path is an exchange in the energy spectrum.

Eternity: All events on an infinite time line.

Event: A point of existence is a unique event on a time line.

Expansion: An increase in distance on a spiral path.

Inertia: Transfers from any changes in the axis of direction's spiral path.

Inference: The act of generalizing from statistical sampling data.

Intersect: An intersect of a time line is a unique piercing or the dividing of a time line segment by a spiral path segment.

Intersection: An intersection is occupied by a unique point of existence on a spiral path intersect with a time line segment. This unique point of existence on a spiral path is a single unique event on the time line segment.

Intervals: The distance on a spiral path segment between each of the events they create on a time line.

Minutia: A point of existence on a spiral path that is also a unique event on a time line and is not observable.

Minutiae: The cohorts of points of existence on spiral paths and their unique events on parallel time lines, which are only observable as a single event in the relative world. This is reflected in quantum theory where complete separation of events is not possible so the terms 'interaction' and 'separation' are used together with the fact of their limitation.

Non-existence: The space on a spiral path not occupied by existence.

Orientation: Orientation is determined by the dominant axis of direction of the universe and provides the topography of the universe from creation to annihilation and recreation.

Parnassus effect: The bulge at the equator of celestial bodies that act as converging meniscus lens. These lens transfer radiating gravity from the core of each respective body to a focal point on a plane extending from the line of the equator.

Path: The distance between any two points on unique spiral segments.

Random: All time lines are random and therefore all events in our universe are random.

Relative world: The relative world is measurable in common terms. What cannot be observed directly can be described in comparative common terms that are based on statistically projected outcomes or events that can be confirmed by the scientific method.

Spiral path segment: The spiral path is infinite and therefore only a segment can be examined. The spiral path follows its axis of direction.

Steady state: A spiral path segment in a universe that is inherently constant and lacks variability. However, variability influences all spiral paths over infinity.

Strata: The stratification of cohorts in multiple spiral paths that have a common dominant axis of direction.

Symbol Δ: This symbol represents the deemed angle that any two time line segments cross but do not intersect.

Time line segment: The time line is straight, fixed, and infinite and, therefore, only a segment can be examined.

Topography: A point of existence has no topography, but its cohorts are the source of the minutiae for the science of topography.

Variability: Variability is where there are constant changes in a spiral path segment. Change is a constant because of variability in the dominant spiral path segment.

Velocity: A change in a spiral path segment.

Introduction

Philosophically, warmth is fundamental to human existence starting from the womb and delivery to the first embrace of our mother's arms. We are drawn to the warmth we seek in childhood and adult relationships: a warm personality, greeting, handshake, hug or kiss. Our life exists in this very narrow, warm middle of the energy spectrum. The Universal Theorem (TUT) is a philosophical exploration of this spectrum.

The mathematical relationships developed by the ancient Greek philosophers Pythagoras (570 BC–495 BC), Euclid (325 BC–265 BC) and Archimedes (287 BC–212 BC) fascinated me as a child and young adult. My interest was piqued again years later during my training as a Certified Internal Auditor. My mentor, Lawrence Sawyer, paraphrased Archimedes', 'Give me a place to stand on, and I will move the Earth', into, 'Give him the standards and he could audit the universe'. Both are simple, but profound, statements and inspired me to broadly research literature on the general theories of relativity and quantum mechanics. The material presented some universal standards but also created voids of uncertainty to explain the largest and smallest events to complete a universal theorem. As an auditor, the non-existence of standards in any business process is very important to me, and I am obligated to facilitate the creation of these necessary standards. It is this auditor mindset that drives my TUT instincts.

Lawrence Sawyer also taught the key to reaching the truth, according to Sherlock Holmes, is to separate the truths from the untruths. To reach my theorem, I will separate existence from non-existence in search of the truth. In my theorem, it is non-existence that, in a literal sense, propels the universe and is one of the concepts that TUT explores.

From a personal perspective, we appear to exist, as humans, in the centre of the visible universe between the infinitely large and the infinitely small. Philosophically, this is a unique position in the scale of our universe, but our observations in either direction of this scale generally rely on tools such as the telescope or the microscope. Through these tools we have discovered both sides of the observable scale in which we exist are occupied. Existence in TUT is based on unique single points of intersections in time and space where there is no overlap, the points are not physically observable and have neither mass nor dimension. But points of existence are the sole source for what is observable, and our relative existence in the physical world in which we live. Existence in the relative world, and to a limited extent the universe, is observable and measurable. Philosophically, relativity is based on measurable ends whereas TUT is based on non-measurable means.

Points of non-existence allow the flow and ebb of the universe moving from one state of matter to another state. The big bang is part of the flow and ebb in which all observable galaxies exist. Like points of existence, it is not measurable. In relative terms it allows the physical world concept of on or off. Off would be the state of non-existence in relative terms. In TUT all points of existence move to points of non-existence. A universal cascade of movement results in annihilation. All movement in the relative world is caused by this TUT proposition.

In the auditing profession, we apply probability tables for selecting transactions to test against our audit attributes. The probabilities of revealing errors in transactions increase as the inherent characteristics of the audit universe are observed and known. TUT is an examination of the universe to primarily identify the attributes of existence, and to a lesser extent non-existence, to help fill the knowledge voids in our relative worlds and reach a universal theorem. My belief is that the truth is like an asymptote whereby we get closer and closer to the truth but absolute certainty, based on probabilities and the history of science found in our relative worlds, is evasive. To eliminate probabilities found in our relative world approach to science, I concluded that, philosophically, TUT must have attributes that are based on absolutes.

In my auditing career, we were encouraged to find solutions sometimes *outside the box*. TUT is an attempt to think 'outside the sphere' of the scientific method successfully applied to our observable and measurable relative existence. The method of my research is limited to the use of authoritative sources that link the philosophical arguments for TUT in relative terms. The pitfall of my approach would be to provide only those relative applications that support TUT, and every effort has been taken to ensure other perspectives have been considered. Since the body of my composition represents my personal views, opinions and philosophies, I have adopted an essay format.

We generally accept in our relative world that the only absolute is *change*. TUT asks the question if this change ultimately leads to annihilation, then 'why'? To address this question, I have developed six TUT absolutes and their related propositions, and, in some, I have also provided corollaries. The linkage of TUT to our relative world is then dealt with in a philosophical context, and finally, in a limited way, to the relative world we live in and of which we are an integral part.

The Six Absolutes and Respective Propositions

1. Space

Propositions:

1.1 *Distance is a spiral segment*

Because in absolute space travelling between two spiral paths:

- Where point **a** in *Figure 1* is on a unique spiral path to point $\mathbf{a_1}$ in absolute space, and point **b** is on a mutually unique spiral path to point $\mathbf{b_1}$ in absolute space, then the locus of distance between the two unique spiral paths is the spiral path segment **a** to $\mathbf{b_1}$

Figure 1

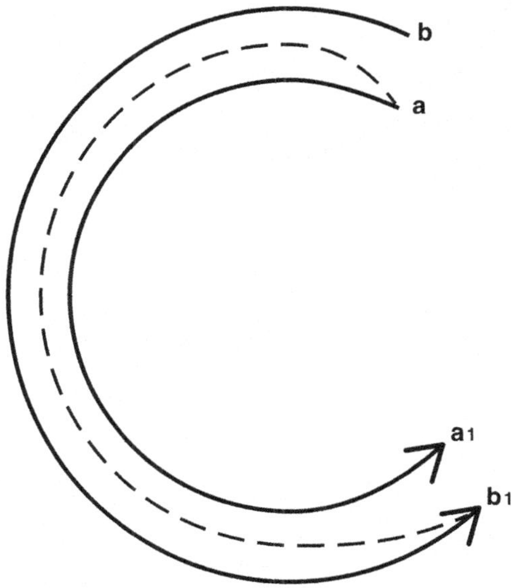

1.2 *The shortest distance between two points is a spiral segment*

Because in absolute space:

- The path of point **a** in *Figure 2* is curved. Distance is the spiral path **a** to $\mathbf{a_1}$
- To arrive at $\mathbf{b_1}$ from **a** the locus of the path in *Figure 3* is a variable spiral segment whose shortest distance is **a** to $\mathbf{b_1}$

Figure 2

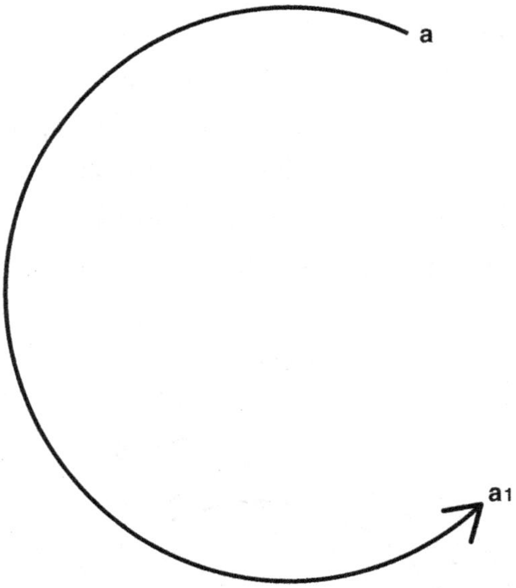

Figure 3

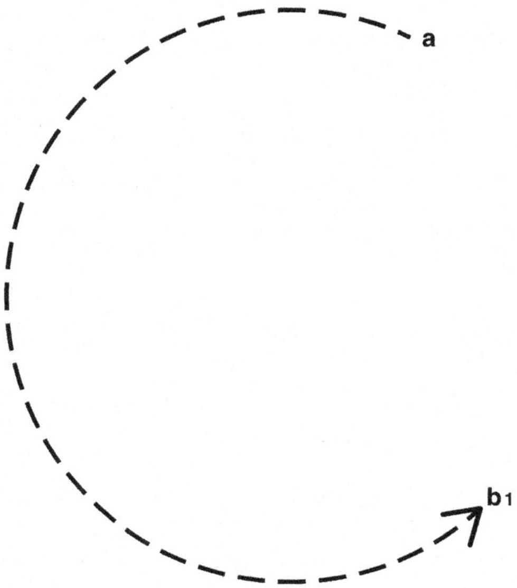

COROLLARIES:

- Distance is between points on any spiral path
- A path is the locus of distance on a spiral segment
- The shortest distance on a spiral segment is the locus of its path

1.3 *The expansion of the universe is a spiral path segment*

Because distance is a curved spiral segment:

- The distance between galaxies in *Figure 4* is the variable spiral path segment **s** to s_∞
- Galaxy g_2 is gaining distance from g_1 and g_3 is gaining distance from g_2 on their variable spiral path segment
- The expansion continues to g_n

Figure 4

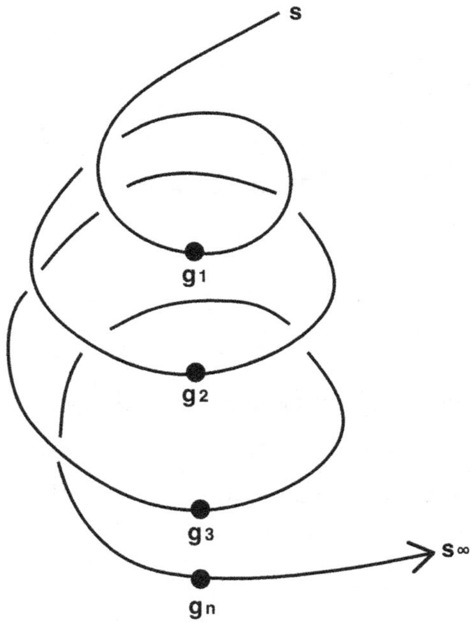

2. Time

PROPOSITIONS:

2.1 *Time is an infinite fixed straight line*

Time is linear with neither a beginning nor an end. A time line transcends all universes and is unique:

- Events **e**, e_1, e_2, e_3 and e_4 in *Figure 5* share the same unique time line segment **t** to t_∞
- Since distance is curved and time is linear, **e** to e_4 cannot be measured

Figure 5

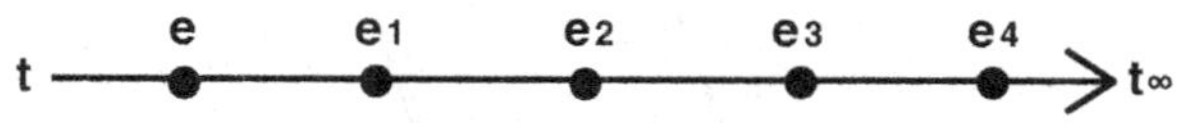

2.2 *Time is only intersected by spiral segments*

Because a time line is fixed and spiral paths are variable, it is the spiral paths that intersect time:

- Since time lines are unique, the spiral path intersections are on an infinite number of time lines
- Spiral segment **s** to s_∞ in *Figure 6* intersects time line segments **t**, t_1, t_2 and t_3 at **i**, i_1, i_2 and i_3 respectively
- Unique time lines are parallel to each other on an infinite number of fixed angles that do not intersect. In *Figure 7* parallel time lines **tx** and tx_1 cross **ty** and ty_1 at angles

Figure 6

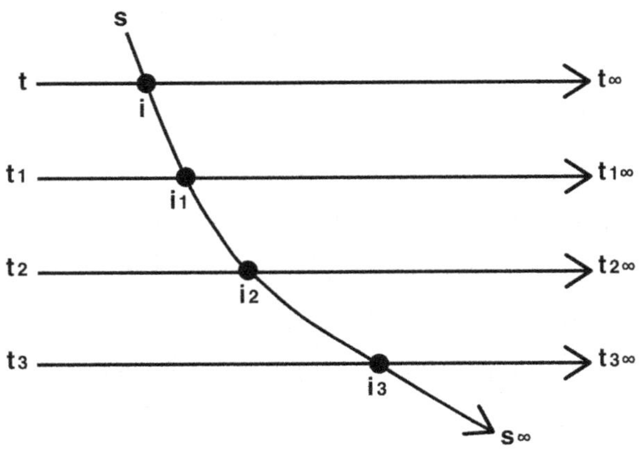

Figure 7

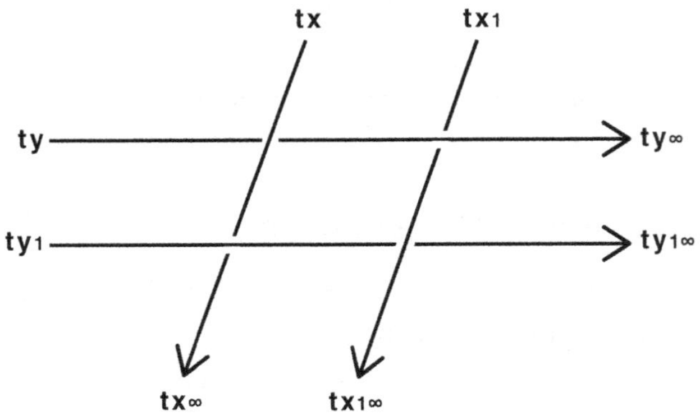

2.3 *Spiral intersects are random events*

Because time lines are infinite parallel lines that cross at infinite angles, the spiral intersects of time lines are random:

- In *Figure 8* time line segments t to t_4 do not intersect
- The time line segments t_2 to $t_{2\infty}$ and t_3 to $t_{3\infty}$ are parallel and do not intersect
- Therefore there are an infinite number of parallel time lines that do not intersect
- The angles that parallel time lines cross other parallel time lines are infinite
- Since the angles are infinite, the spiral segment intersects of time lines are random events

Figure 8

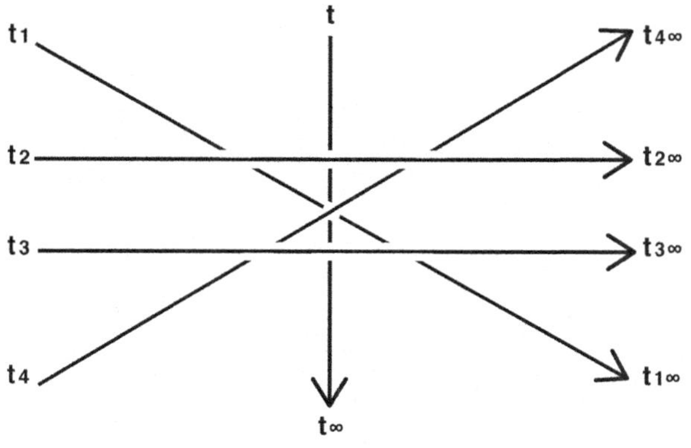

3. Existence

PROPOSITIONS:

3.1 *Events on a time line are points of existence*

Because it is a spiral path that intersects a time line:

- An event **e** is created by a spiral path at the intersect
- In *Figure* 9 a point of existence is the **e** intersect on time line segment **t** to **t∞** by a spiral path segment **s** to **s∞**

Figure 9

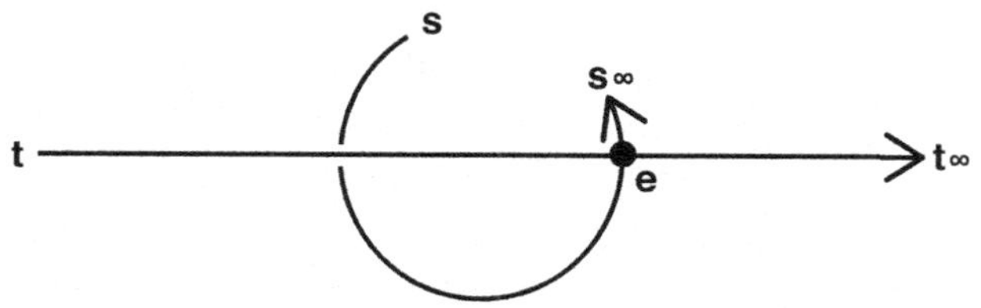

3.2 *Existence occupies a unique point of intersection*

Because the spiral path segment intersects the time line segment once at **e** on *Figure* 9, existence occupies a unique point of intersection

COROLLARIES:

- The intersection is a singular occupation of space in time
- An infinite number of singular mutual occupations in time lines by spiral paths is universal existence

3.3 *No two points of existence occupy the same point of intersection*

Because the point of existence is a unique event on a time line segment and distance is a curved path:

- All intersects with a time line segment by spiral path segments are separated by their variable paths on the time line that is immutably fixed. The variable spiral paths **s** to **s∞** and **s1** to **s1∞** on *Figure 10* will only intersect with time line segment **t** to **t∞** at **e** and **e1**
- These two spiral path segments are unique. Their single event intersects on time line **t** to **t∞** are occupied by single points of existence

Figure 10

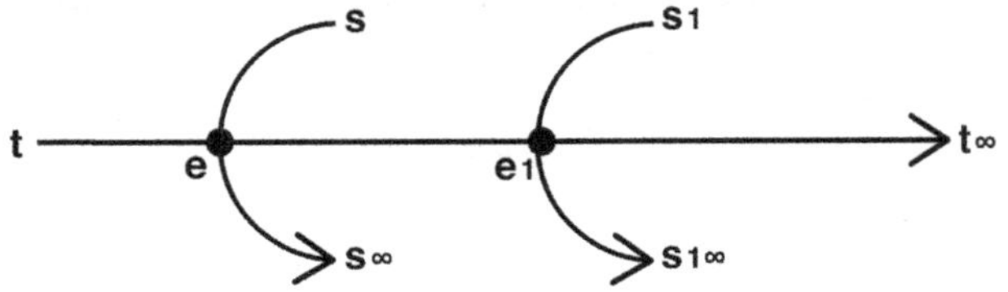

4. Non-Existence

PROPOSITIONS:

4.1 *Points of non-existence occur on spiral path segments between all points of existence*

Because existence is at the intersection of a spiral segment with a time line segment:

- A non-intersect of a spiral segment with a time line segment is non-existence
- The points of non-existence are between the random intersects of spiral segments with a time line. In *Figure 11* intersects of spiral segments **s**, s_2, and s_4 are points of existence on time line segment **t** to $\mathbf{t_\infty}$ at **e**, $\mathbf{e_1}$ and $\mathbf{e_2}$
- The non-intersects of spiral segments $\mathbf{s_1}$ and s_3 on time line **t** to $\mathbf{t_\infty}$ are points of non-existence **ne** and $\mathbf{ne_1}$

Figure 11

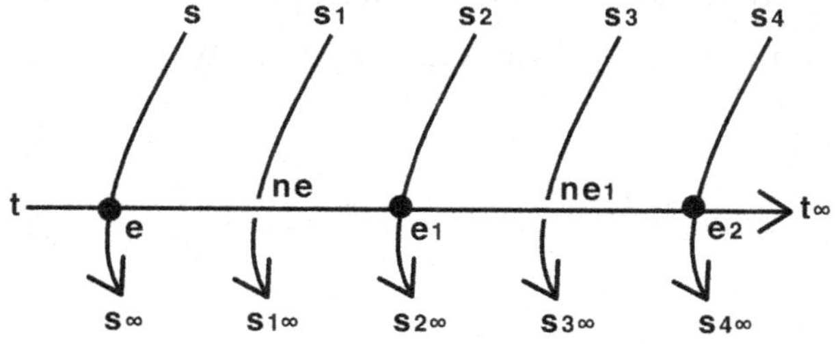

4.2 *All points of existence move to points of non-existence*

Because all variable spiral path segments intersect on random time line segments:

- Existence is variable and random
- The movement from existence to non-existence is caused by variations in each unique spiral segment path that changes them from an intersect on one time line segment to a non-intersect on another unique time line segment and back to an intersect on another unique time line
- Variability of existence and non-existence is directly correlated to the variability of a spiral segment path. On the spiral segment path **s** to **s**∞ in *Figure 12*, there is:
    - ➤ Existence at time line segment **tx** to **tx**∞ at **e**
    - ➤ Non-existence at time line segment **ty** to **ty**∞ at **ne**
    - ➤ Existence at time line segment **tx**1 to **tx**1∞ at **e**1
    - ➤ Non-existence at time line segment **ty**1 to **ty**1∞ at **ne**1
    - ➤ Existence at time line segment **tx**2 to **tx**2∞ at **e**2
    - ➤ Non-existence at time line segment **ty**2 to **ty**2∞ at **ne**2
    - ➤ Existence at time line segment **tx**3 to **tx**3∞ at **e**3
    - ➤ Non-existence at time line segment **ty**3 to **ty**3∞ at **ne**3

Figure 12

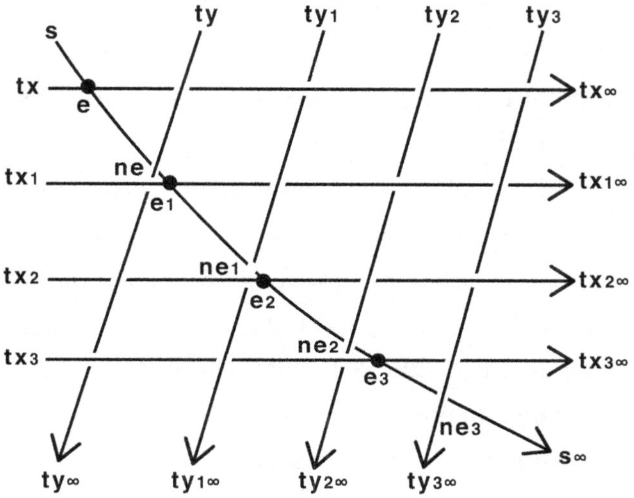

Corollaries:

- The less variable the spiral path segment is the more points of existence
- The more variable the spiral path segment is the more points of non-existence

4.3 *Motion is a change in a spiral path*

Because the variation in the curve of a spiral path segment is movement between existence and non-existence:

- Movement increases in proportion to variation in spiral segment paths
- The less variation in spiral path segments the more consistent the movement through time line intersections
- The more consistent the movement through time line segments, the more aligned the spiral path segment becomes with parallel time line segments. The spiral path segment **s** to **s**$_\infty$ in *Figure 13* consistently intersect at points of existence **e** to **e**$_4$ with parallel time line segments **tx** to **tx**$_4$ respectively. The spiral path segment **s** to **s**$_\infty$ consistently does not intersect time line segments **ty** to **ty**$_3$ at **ne** to **ne**$_3$. Therefore, the points of existence **e** to **e**$_4$ are aligned with parallel time line segments **tx** to **tx**$_4$

Figure 13

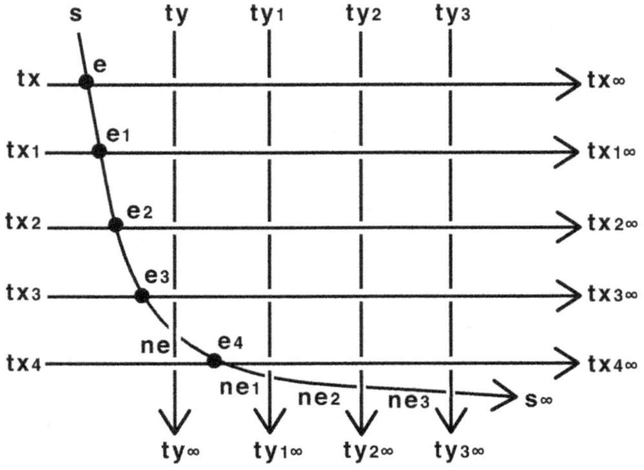

Corollaries:

- Movement decreases where the spiral path segment is aligned with parallel time lines
- Spiral path segments in a steady state contain greater points of existence
- Since all time line segments on an infinite number of angles share an infinite number of parallel time line segments, steady states of existence are throughout the universe
- Conversely a spiral path segment containing a larger portion of non-existence will attract existence from steady state spiral paths

5. Energy

PROPOSITIONS:

5.1 *Variability in a spiral path segment is energy*

Where spiral path segments are variable:

- Energy is a change in the path of a spiral segment
- In *Figure 14*, starting at point s, changes on projected paths **sp** to **sp**1, **sp**2, and **spn** the energy exchanges are at **e** to **en** on spiral path segment **s** to **s**∞

Figure 14

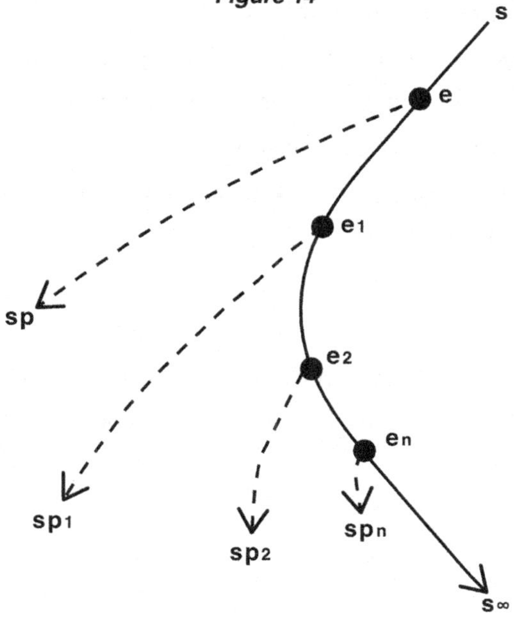

CORROLARIES:

- The intensity of energy is determined by variability
- Since **e** to **e**n is on a spiral path segment, then **e** equals **d** (distance)
- As **d** decreases, energy is absorbed and conversely as **d** increases, energy is released

5.2 *Tightening of a spiral path absorbs energy*

Because there are high levels of existence in a tight spiral segment path:

- Energy is the exchange between points of existence and non-existence in spiral paths
- And where the points of non-existence are eliminated, energy is intensive and stable

COROLLARY:

- Energy is conserved in a steady state of high energy in a tight spiral path segment

5.3 *Expansion of a spiral path releases energy*

Because energy is conserved in a tight spiral path segment:

- Energy is released as existence on a variable spiral path segment moves to non-existence
- The change from existence to non-existence expands the spiral path segment

COROLLARY:

- All movement is between existence and non-existence and therefore is the source of energy exchanges

6. Orientation

PROPOSITIONS:

6.1 *The demarcation of orientation of a spiral path segment is determined by its axis of direction*

Because a spiral path segment reverses around its axis of direction:

- In *Figure 15*, the demarcation of orientation on the spiral path segment s to s_∞ is from its axis of direction **a** to $\mathbf{a}_\infty$
- At points **o** to $\mathbf{o}_n$ on spiral segment s to s_∞ the orientation is to their axis of direction **a** to $\mathbf{a}_\infty$

Figure 15

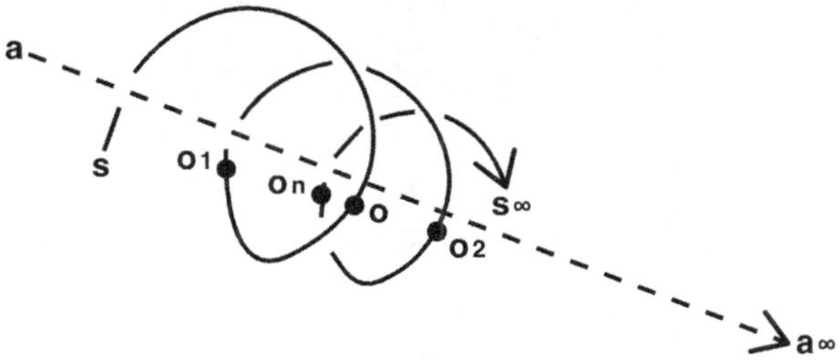

6.2 *Orientation is on the axis of spiral path segments*

Because all space is curved:

- In *Figure 16*, spiral segment path s to s_∞ axis of direction **a** to $\mathbf{a}_\infty$ is also oriented to its dominant axis of direction $\mathbf{a}_1$ to $\mathbf{a}_{1\infty}$
- The axis of direction of **a** to $\mathbf{a}_\infty$ and $\mathbf{a}_1$ to $\mathbf{a}_{1\infty}$ are also spiral path segments
- The axis of a dominant spiral path segment creates orientation between spiral path segments

Figure 16

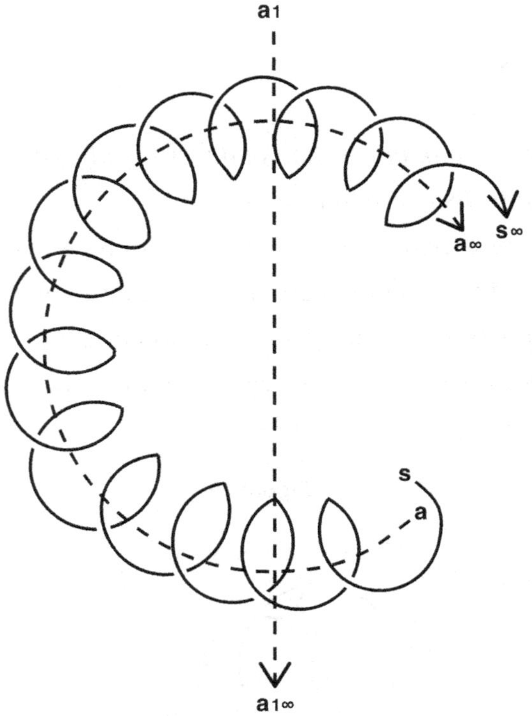

Corollaries:

- The axis of spiral path segments are spiral path segments
- Dominant spiral segments are expansive and variable
- Orientation is determined by the axis of dominant spiral path segments

6.3 *Orientation determines the topography of existence*

Because orientation is determined by the dominant axis of direction of a spiral path segment:

- The variability of the orientation on the axis of direction is the source of expansion
- Where the axis of orientation has a very tight spiral path segment, with high points of existence in a steady state, the topography moves outward from the dominant spiral path to points of non-existence
- The outward movement creates a cascade of energy transfer between existence and non-existence

COROLLARIES:

- Since existence moves to non-existence expansion ceases at a state of equilibrium,
- Equilibrium eliminates stability in steady state dominant spiral paths thereby changing the topography of universal existence

The Application of TUT in Relative Terms

The Philosophical Context

In our physical world, the relative is relevant and thus to make TUT relevant, we must speak and write in relative terms. But using common terms can also trap us in words that imprint on our minds images of our relative world. For example, the words sunrise and sunset trap us in the relative world of myths. They perpetuate an archaic notion of planetary motion and our place in the solar system. In TUT the sun neither rises nor sets.

Culture in our relative world is based on contrasts. Art, in particular, is all about contrast, and contrast is the basis for both our pure and applied sciences. Music is a contrast in sounds, tempo and melody. The painter uses colour, texture and shape for contrast. For writers and poets, we convey the images and the emotions from the words we use as contrast. Science is the use of scale and boundaries whether it is time, distance, temperature or mass. Relative laws of physics, which are also applicable to arts, are developed from these contrasts. Both art and science are intricate components of the energy spectrum explored by TUT. The energy spectrum in TUT is the means and the contrasts found in the relative world are the ends. The ends are not only familiar in our relative world but necessary to our biological comfort, safety, mental health and existence.

Being part of the relevant world for so long, I had to think *outside the box* if the results were to be applicable universally. However, I soon discovered the irony of this approach because in TUT there are no boxes. The Euclidian geometry to create a box does not exist in TUT because space consists of curved spiral path segments and therefore no Euclidian geometry exists. For me to get out of the box, the adoption of the spiral segment path for TUT became logical and absolute. Because the spiral path is based on its axis of direction and both are infinite, I adopted the use of segments to establish boundaries. Having done this the same approach was used for time lines.

This led to examining, philosophically, the virtual image we create when we draw a box or when we think of a box. In the nanosecond the pen touches the paper in the relative world, the locus of the ink from the pen point's spiral path through absolute space becomes variable as it dissipates energy at every point of non-existence. A TUT proposition is that energy is a change in a spiral path and where every point of existence on random time line intersects. The same is true for the thought of a box and the related mental picture of the box retrieved by our minds. The nanoseconds of energy that the intensive, creative thought processes requires to have the image prepared would also be dissipated in absolute space. Our piece of paper and our mind would still retain their transferred virtual images in high energy steady state until dissipating their energy into points of non-existence, according to my TUT propositions. Therefore, neither pieces of paper nor our minds endure in our relative world.

Philosophically, TUT must go beyond the attributes of our personal existence to the attributes that are universal. It surprised me in developing TUT that, at some point, the philosophy of science and theology become congruent in the big picture. Both, in certain respects, require faith in the unseen. Both also have leaders that prophesize future outcomes based on their faith in the unseen. In the Bible, St Peter wrote, almost two thousand years ago that in the end, *'... the heavens will pass away with a loud noise, and the elements will be dissolved with fire ... we wait for new heavens and a new earth ...'* (2 Peter 3.8-14).

If we follow the time lines from our personal points of existence to infinity, then according to the TUT propositions in 6.3, the changes in topography of celestial bodies, including earth, are shared on the same parallel time lines. These time lines are also shared with other universes that are in various stages between annihilation and reconstruction. The death of our relative world will give birth to a new relative world, including all of the same challenges we currently face. In other words, physical and mental suffering will always be part of our biological existence in every reconstruction phase because the cohorts of our points of existence are common and shared universally. My philosophical conclusion is whatever was, is!

Pure Perfection

We are in detection of the world's imperfection,
Our sunsets having little symmetry or geometry,
The chaotic ocean waves crashing and splashing,
Winds from all quarters whirling and swirling,
The random rain pitter-pattering,
The planet earth being neither round nor sound,
A human race always resistant to being consistent,
And a god who plants weeds amongst our seeds.
Yet, the sum of all imperfection… is pure perfection!

—*Opa's Journey…*

CONVERSION

TUT propositions describe transfers of existence to non-existence. These transfers represent energy conversions. The fundamental and absolute conversions include the entire energy spectrum.

In the relative world, gravity is considered a weak link in respect to the rest of the energy spectrum. In TUT there is a universal oxymoron when examining gravity because it is both weak and strong. It represents a very tiny tight spiral path, but its strength is in its fundamental dominance as the catalyst of all energy conversion.

The energy spectrum consists of spiral paths having high levels of existence. The spiral paths are in steady state and have cohorts on parallel time lines. These cohorts provide the energy spectrum. Energy transfer for each category in the spectrum is initiated by gravity, which is the dominant spiral path. Therefore, gravity is at the centre of all energy conversions and enables the transfer of all other energy in the spectrum.

On earth most of the gravity is transferred to the crust and, secondarily, to the oceans. These constant tiny tight spiral paths are in a steady state and radiate out from the earth's core, according to the TUT propositions. Mountains on the earth's surface receive additional gravity transfers, and the effect is observable on mountain tops, including undersea mountain ranges. The gravity transfer to undersea mountains is also observable whereby the ocean surface directly above the mountain tops rise relative to the surrounding seas that follow the contours of the lower sea floor.

The closer to the earth's core the more frequent the encounters with the transfers of the radiating tiny tight spiral paths of gravity. In other words, this spreading of individual spiral paths provides high levels of energy conversion closest to the source, which is the centre of the earth. The tiny tight spiral paths transferred to the earth's crust, oceans and mountains, which are not fully absorbed radiate outwards in transfers to absolute space.

In *Figure 17*, the individual radiating tiny tight gravity spirals are constant from their origin at g_o to g_n. The transfer of gravity at the object c_1 to c_2 is **g** to g_7 and the transfer at object c_3 to c_4 is g_1 to g_5 where both objects are equal. Therefore, the further an object is from g_o the gravity transfer decreases.

Figure 17

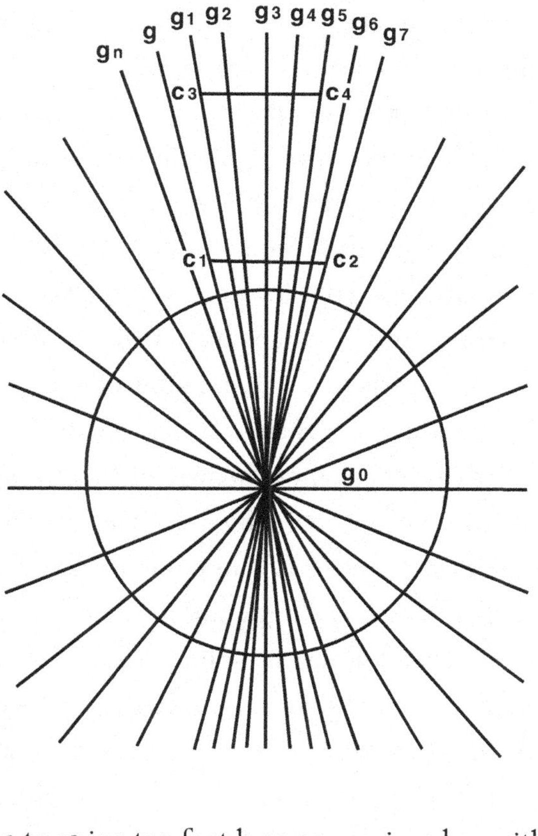

If the object c_1 to c_2 is a ten feet long square iron bar with dimensions of 3 inches by 3 inches, the gravity transfer through the bar would be less on the top surface of the bar than received from g_o on the bottom surface. If the bar were reoriented having one end of the bar c_2 pointing to g_o, the gravity spiral path encounters would be less frequent on the smaller surface of 3 inches by 3 inches. However, the absorption of gravity transfer over the 10 feet length of the bar would be much greater thereby reducing the gravity transfer on the surface of **c**$_1$. The total gravity transfer is equal in the metal bar regardless of its orientation to **g**$_0$. The same would be true if the bar was given any other shape, including a sphere, because according to TUT propositions, all unique tiny, tight gravity spirals are constant. The change in absorption in this example is based on **d** (distance) that, according to TUT propositions, is equal to **e** (energy) where the surfaces of the bar are not straight but curved spiral path segments.

Thermodynamic processes allow cloud masses to store large reserves of energy. The core of the inherent energy spectrum, according to TUT propositions, is gravity. Since gravity consists of tight spiral paths, the transfer of gravity from the concave shape of the bottom of large dense clouds, facing the convex surface of the earth, is focused on the earth's surface rather than being radiated outwards. This focus becomes the axis of direction for the spin of the cloud mass and creates a source of gravity exchange that is greater than the gravity from the earth's surface. The more benign exchanges are the updrafts created in a storm's shifting core. The most dangerous are the intense energy transfers at the centre of the clouds' axis of direction that contact the earth's surface and draw surface materials away from the earth. The axis of direction for the spiral path of the energy transfer follows the uneven topography of the earth from the movements of the clouds shifting core until the cloud mass dissipates its energy reserves. When the frequency of the energy transfers from the clouds becomes less than the radiating energy transfers from earth, the earth's gravity transfers become dominant. Any surface materials caught by the clouds' axis of direction are then released to earth.

The same processes exist on the earth's sun whereby very large clouds of extremely hot gasses create focused gravity exchanges on their concave side with the sun's radiant convex surface. The solar flares are the dissipation of the energy spectrum drawn up by the contact of the axis of direction, from the shifting core of the immense hot gas and gravity laden clouds, focused on the sun's surface.

The gravity at the core of the sun, other stars, planets and moons create their inherently individual celestial spheres. The extreme gravity transfers reaching the surface of the stars create the stars' respective individual energy spectrums over the surface by providing an axis of direction for points of existence on an infinite number of variable spiral paths. Because of the great transfer of gravity on the surface of the stars, the gravity transfer creating the energy spectrum prevents some of the spectrum from totally escaping the surface. However, a neutron star with only its core remaining has had its energy spectrum primarily reduced to gravity. This large and direct transfer of gravity shrinks the core of the neutron star as it dissipates its energy.

Gravity also provides the axis of direction for atoms. Their nucleus consists of gravity. The atoms fundamental structure consists of electrons in unique spiral paths around the nucleus that provides their axis of direction. The orbit of the electrons' spiral path around their respective nucleus is where the dominant spiral path determines their position in the energy spectrum. Each electron is like a tiny, spinning, electrical charge and each amounts to a tiny magnet. Where the four outer electrons of iron atoms are lined up, they create a magnetic field in a bar of iron. A large magnetic field around the earth is also induced by the electrical currents created by the earth's spinning metal core. The following matrix in *Figure 18* provides examples of some of the orbits in the energy spectrum. The tightest orbit has the highest energy.

Figure 18

Energy Spectrum	Orbit of Electrons Around the Nucleus
X-ray	Out of orbit
Ultra-violet	Almost out of orbit
Light	Close to nucleus
Infra-red	Closest to nucleus

Energy conversion takes place when the orbital spiral paths of points of existence of electrons, change their orbit around their respective nucleus. It is the gravity transfer from the nucleus that attaches to the energy spectrum. Because **d** (distance) = **e** (energy) the change in the orbit = **d** = **e,** according to TUT propositions.

Large energy exchanges initiate from the centre of the source outward in an infinite number of angles. An observation of a distant super-nova from earth to determine its topography would only measure those energy exchanges intersecting our shared parallel time lines. The bulk of the energy exchanges would be on time lines not shared by earth. These exchanges would radiate outward, according to TUT, and connect with other universes not mutually sharing our universe's parallel time lines.

Figure 19

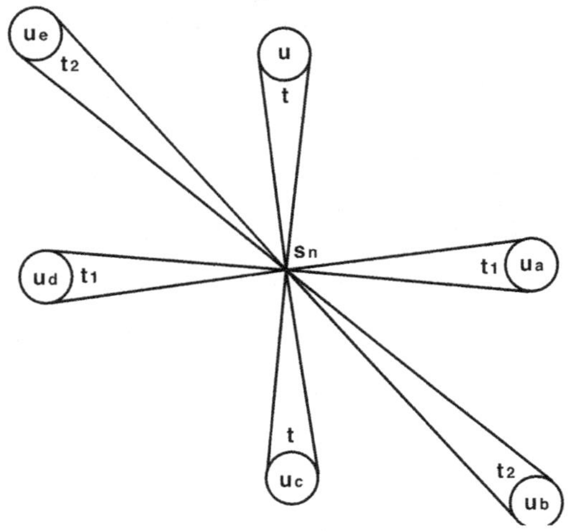

In the foregoing example, *Figure 19,* the super-nova s_n energy transfer **t** only exists within the range parallel time lines mutually shared by earth's universe **u** and universe u_c. Earth's universe **u** does not share the unique parallel times lines t_1 that are mutually shared by universes u_a and u_d. The same would hold true for parallel time lines t_2 that are mutually shared by universes u_b and u_e. Time lines are unique, fixed, and infinite and cross each other without intersecting on an infinite number of angles and, therefore, the number of universes sharing energy transfers from sn on parallel time lines is also infinite. However, the intensity of energy transfer diminishes as the energy radiates to the most distant universes.

Unlike the Roman god Janus, we can only see in one direction. To make an observation of transfers of the energy spectrum in absolute space, our vision is confined by the parallel time lines of the observation. Any change in the direction of the observation moves our vision to different sets of parallel time lines that, according to TUT propositions, cross each other at an infinite number of angles without intersecting. This provides a multi-dimensional view of the universal energy transfers as our eyes shift between the angles inherent in parallel time lines. Each observation, however, is unique to the observer. No two or more observers share exactly the same observation even if recorded on cameras. It is analogous to the testimony given by witnesses to an automobile accident or a crime scene. The sworn testimony of the witnesses can often appear contradictory.

In the relative world, each visual observation on a set of parallel time lines appears as a single event in the energy spectrum transfer. However, using a high speed motion camera, we observe a virtual series of events that visually is one event. According to TUT propositions, events on time lines are unique single intersects on time lines. The observations in the relative world are limited to the use of blunt instruments that require massive intersects on parallel time lines for observation. The application of the Werner Heisenberg uncertainty principle applies to the limitation of these blunt instruments. The singular unique events defined by TUT are not observable because they happen before they can be observed. Philosophically, it comes to the relative world question of how many hairs on a man's chin make a beard? Does a singular hair, or two or three hairs, on a man's chin make a beard? TUT propositions are absolutes and the relative world relies on quantifiable ranges for describing their observations.

The unique tiny, tight spirals of the energy spectrum, defined by TUT propositions, are more readily transferred by surfaces in the relative world that have an abundance of points of non-existence. Gas has the most points of non-existence for energy transfer followed by liquids. Crystal has the least points of non existence, and where the surface is finely sealed by polishing, it reflects a significant portion of the energy spectrum. The individual tight energy spirals arrive on earth at the perpendicular angles created at their source. The reflection of the energy spectrum in lens or dishes that facilitate refocusing their perpendicular angles from a multitude of parallel time lines is a multi-dimensional tool for observation of our universal topography. However, the energy spectrum radiating from distant celestial bodies in spiral paths intersects a multitude of parallel time lines from an infinite number of angles. The parallel time lines that connect the observer and the energy spectrum transfers from a great distance reveal the source of the transfers but not the position of the source relative to the position of the observer. It is analogous to seeing the source reflected off a mirror because the frequency of time line intersects by the energy spirals change our observation to the final parallel time lines that connect our observation to the energy exchanges leaving the surface of the mirror. The axis of direction from the energy sources that are not absorbed by the mirror, as an energy spectrum transfer, are reversed in the reflection from the surface of the mirror.

The mechanics of this energy spectrum reversal and transfer are illustrated in subsequent discussions.

The TUT philosophy has identified itself as the universal means and the relative world presents the practical ends in which we live. An example that captures the essence of this philosophy is the movement of celestial bodies in absolute space. As they move closer, encounters of the energy spectrum, according to TUT propositions, become more frequent and, conversely, there are less encounters of the energy spectrum as the bodies move apart. This creates an apparent shift in the energy spectrum that is useful to measure movements of celestial bodies. The measurements, named the Doppler effect, are based on relative calibrations of the red-shift in the energy spectrum.

Similar relative calibrations were developed from measuring energy spectrum emissions of radio signals from the Geosal satellite over the earth for Doppler shifting as it passed overhead. The conclusion was that a slower relative speed meant a higher orbit and, conversely, a higher speed meant a lower orbit. In these examples, TUT defines the means and the relative Doppler calibrations provided the practical ends for application in the relative world that we live in.

According to TUT propositions, time is not measurable and, therefore, intervals are determined on the spiral path segments as the distance between events on a time line. The tighter the spiral path segments are the shorter the distance between intervals.

Figure 20

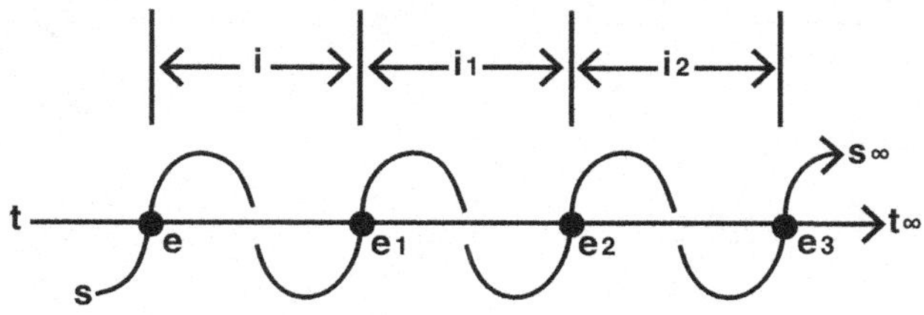

In *Figure 20*:

- **i** – interval
- **t** – time line segment
- **e** – event
- **s** – spiral path segment

The intervals **i** to **i**2 on time line segment **t** to **t**∞ are determined by each unique intersect by spiral path segment **s** to **s**∞. Therefore:

- Eternity is all events on an infinite time line
- The intervals between events are determined from their spiral path intersects
- Intervals are measured between points of existence on their spiral path intersects with time
- Intervals have a direct correlation with a change in their spiral path segment intersects

Corollary:

The frequency of intervals on a time line is directly proportionate to velocity

Figure 21

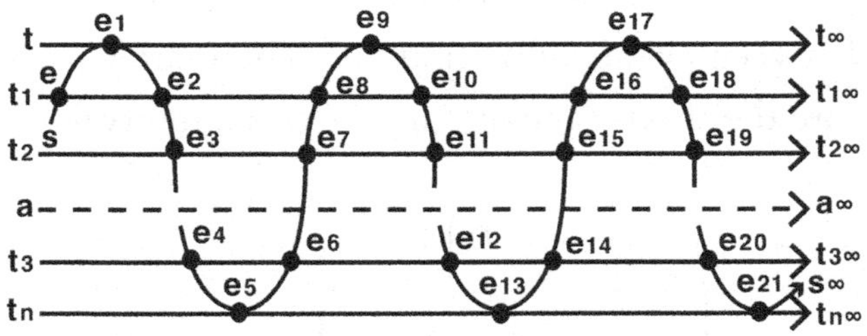

In *Figure 21* above:

- **t** – time line segment
- **e** – event
- **s** – spiral path segment
- **a** – axis of direction

Intervals on parallel time lines **t** to $\mathbf{t}_n$ are between points of existence at each event **e** to $\mathbf{e}_{21}$ on the spiral path segment **s** to $\mathbf{s}_\infty$. The intervals between events on parallel time lines are distances on the spiral path between each event. These intervals shorten on tight spiral paths and increase as the energy spectrum is transferred outwards from the spiral segment's axis of direction **a** to $\mathbf{a}_\infty$. Where intervals are large, the energy transfer is low (e.g. radio frequencies). In stable, tight spiral path segments, with small intervals, the energy spectrum transfers are consistent and predictable (e.g. the electropositive element cesium).

Radio telescopes collect and refocus the intervals of the energy spectrum transfers for relative calibrations and measurement. The frequencies of intervals are perceived by the observer as waves. All points of existence on a unique spiral path are intersects with unique parallel time lines. For this reason, large spiral path segments have greater distances between intervals from an infinite number of intersects with parallel time line segments. Therefore, radio frequencies require a large receptor to increase the frequency of detecting and recording the intervals between events on the parallel time lines.

Figure 22

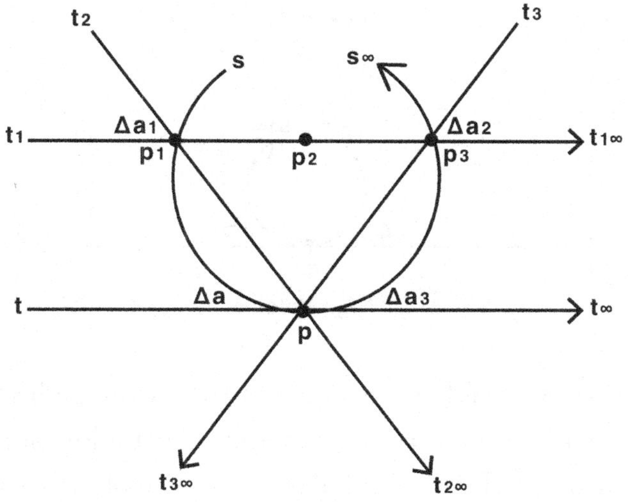

In TUT velocity is the angle at the intersection of a spiral segment to a time line. TUT propositions have been applied in *Figure 22* to space that is a curved spiral path segment and time line segments that are straight and cross each other, without intersecting, at an infinite number of angles.

In *Figure 22* spiral path segment **s** to **s_∞** intersects time line **t** to **t_∞** at point of existence **p**. Parallel time line **t_1** to **$t_{1\infty}$** crosses but does not intersect at points of existence cohorts **p_1**, **p_2** and **p_3**. Point of existence **p_2** is on a unique intersect on the centre axis of direction spiral path but with another time line.

Time lines **t_2** to **$t_{2\infty}$** and **t_3** to **$t_{3\infty}$** cross, but do not intersect time lines **t** to **t_∞** and **t_1** to **$t_{1\infty}$** at **p**, **p_1**, and **p_3**, at angles Δa, Δa_1, Δa_2 and Δa_3. Where velocity is constant in *Figure 22* angles $\Delta a = \Delta a_1 = \Delta a_2 = \Delta a_3$. Any variation in these angles represents a change in velocity as the spiral path s to s_∞ tightens or loosens from energy spectrum exchanges.

Figure 23

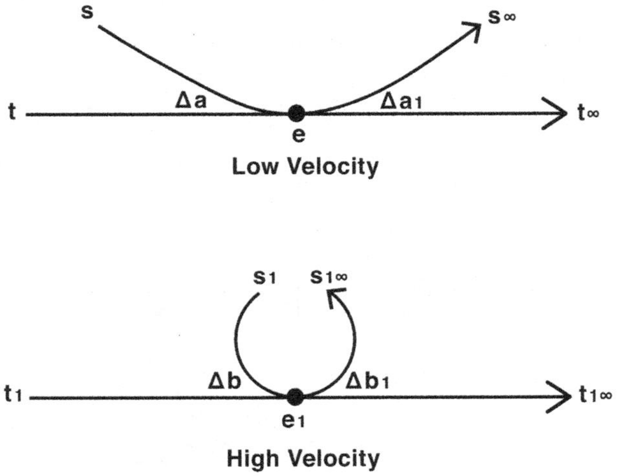

In *Figure 23* velocity is low where the spiral segment path **s** to **s_∞** intersects time line **t** to **t_∞** at point of existence **e** at a low angle **Δa**. The velocity is higher where the spiral segment path of **s_1** to **$s_{1\infty}$** intersects time line t_1 to $t_{1\infty}$ at a point of existence e_1 at a greater angle Δb. Where the spiral segment path s to s_∞ approaches linearity, it becomes bereft of points of existence and is dominated by points of non-existence. These points of non-existence readily accept energy spectrum transfers from high velocity, high energy and tight spiral paths. The high velocity and tight gravity spirals become the axis of direction for the rapid big bang creation of almost infinite energy spectrum transfers to low velocity spiral path segments where our universe reaches its linearity.

Inertia is the change induced on spiral path segments by energy transfers from their axis of direction. As the spiral segment widens, and velocity decreases, inertia is the energy transfer from the axis of direction thereby weakening the high energy tight spiral path of the dominant axis of direction. The converse is true where velocity increases. As the spiral path segment tightens, the energy transfer of inertia strengthens the high energy tight spiral path of the dominant axis of direction. There are no inertia transfers of energy if the velocity of the spiral path segment is constant.

The most obvious and observable transfer of the energy spectrum is in front of our eyes every morning. Regardless of the shape of our sink, bathtub or toilet, the water spirals down the drain. The energy spectrum, according to TUT propositions, is in the spiral path that follows its axis of direction. The inertia transfer of gravity from the axis of direction is radiated outwards and the water makes this transfer visible. The small inner spiral paths receive the greatest energy transfer thereby holding the water close to the axis of direction. There is less energy transfer to the outer spirals as they move slower. The axis of direction for these outer spirals, within spirals, is according to TUT influenced by the dominant axis of direction at the centre that draws them down the drain. The energy spectrum transfer in liquids constrained in pipes becomes very tight spiral paths as its velocity increases. The tight spiral paths are observed in the liquids as laminated flows rather than turbulent flows.

Each drop of water, in every mighty river on earth that fills the oceans, starts with a single point of existence and so it is with our universe. All celestial bodies start with a single point of existence. There are no building blocks for our universe but rather an awesome cohort of points of existence that challenge our imagination. The metamorphous of the cohorts of existence in each drop of water from gas to liquid to crystal is the result of the same energy spectrum transfers that move our universe from existence by annihilation to an amorphous non-existence and back again to create a new universe.

Rustling Leaves

In the early autumn breeze I have heard
The rustling leaves clinging to a tree
And seen them in utter silence released
As they quickly darted around me.

Some are grasped by the sighing pines,
That only momentarily hold them fast,
Until their swaying swooping boughs
Loosen their grip to free them at last.

Those free, join in an exhilarating dash
To find a resting place on the forest floor,
And begin their long metamorphosis
To become rustling leaves once more.

—*Opa's Journey...*

Biology

The application of TUT to the relative world must also include the multitude of organisms, including ourselves, whose cellular existence has been observed physically and by extension. TUT applies to the infinitely large, small and what is in between. Biology is more in the realm of the infinitely small.

The cohorts of existence follow critical paths similar to the processes of building ships, facilities and houses. The events on each of the critical path time lines are linear by definition and expression. Certain events must occur before other events take place. For example, in simple terms, the foundation of a house is the first event in its construction followed by floor joists, floors, walls and a roof. While these take place, there are parallel events that can, and in some instances must, take place during the construction of the building fabric. This would include, plumbing, glazing, electrical wiring and heating and venting to name a few. The last events are the drywall, trim, paint and floor covering in that linear order. The cohorts of existence follow a universal critical path. They are driven by energy exchanges that also influence the creation of biological cells and structures. The science of cytology, the study of cells, is based on our relative world ends, while TUT is based on the means. The biological critical paths lie in the reproduction structures and processes of male and female. The male follows the female's parallel time lines in their critical paths to reproduce. The female, on the other hand, has a whole series of internal critical path parallel time lines to initiate fertilization, lodging in the uterus, breast development, carrying and birthing of the fetus and subsequent feeding in that order. Each critical path represents a unique exchange of the energy spectrum between different cohorts of existence that share the path as events along parallel time line segments.

The energy spectrum transfers between cohorts of existence in atoms change the atoms from one element to another. Energy spectrum transfers between atoms combine them into molecules. In a steady state of energy transfer, the points of existence stratify, based on their inherent points of non-existence, as gas, liquid or crystal.

The thermodynamic processes of the water molecules in oceans and seas provide the greatest long term steady state within the energy spectrum. Planet earth has provided this energy spectrum, having minimal fluctuations, to atoms and molecules, allowing their replication in narrow bands of the energy spectrum. The replication of molecules, according to TUT propositions, requires points of non-existence for the process. Water is ideal for replication by energy transfer and provides mobility with its points of non-existence. This leads to further stratification where, according to TUT propositions, gravity is at the core of stratification and provides the axis of direction.

The core becomes the nucleus and the molecules on the most outer stratum become the porous exterior of a cell. The porous exterior allows exchanges between cells having similar cohorts. New attributes are created by these exchanges and the attributes are retained by the cells. As the cell retains more energy transfers from other molecules, it reaches a threshold where it is forced to release access energy from the nucleus. The acquisitive nucleus cannot accommodate any further absorption since every part of the energy spectrum transfer contains gravity. The tight spiral path of gravity is unique and cannot be shared so the nucleus splits and sheds its access gravity along with a share of the cell stratum. This replication releases the access energy and the cells once again start the energy spectrum absorption.

The specialization of cells emerges from this basic process by carrying their attributes with them and passing them along through replication. Specialized cells also rely on different bands of the energy spectrum. In the TUT propositions, energy moves outwards and accounts for not only replication but also mobility. The specialized cells replication event takes place only if they have the appropriate level and type of energy spectrum transfer. This is the driving force for the evolution of specialized cells and organisms. For example, in the relative world, muscle cells must be mobile and, therefore, require greater energy transfers to accommodate the expansion and tightening of the variable spiral paths of their points of existence. As energy (heat) is released, another specialized cohort cell transfers the excess energy to the specialized skin cells or organs, for release, use or retention.

The cohorts of existence in specialized cell structures not only require certain energy spectrum transfers for their life cycles but also require the transfers in specific quantities. According to TUT propositions, the energy spectrum radiates from its centre and the frequency of intersects of tight spiral energy transfers increase when approaching the source. Standing near a hot stove on a cool day is usually within our narrow comfort zone. If we get too close to the hot stove, the infra-red radiation would overwhelm our specialized skin cells. The ultra-violet energy transfers to our skin cells are necessary, but if the transfers are either too much or too intense, they would exceed the cells' capacity for energy absorption and transfer. This would result in radiation poisoning. Other energy spectrum transfers such as x-rays penetrate the body; however, in excess or prolonged quantities, they would alter the functions of other specialized cells in all of the body's organs and skeletal structures.

Our bodies also function within a narrow range of gravity. A long period of low or no gravity would slow energy spectrum transfers within the cell structures. Since all energy spectrum transfers contain gravity in their axis of direction, it is necessary for consistent replication within the cells. Large gravity transfers would block the energy spectrum transfers between the cells by reducing their points of non-existence. In either case, all internal cohorts of existence would respectively lose their ability to transfer energy within the cells or between the cells.

The ability to transfer energy within the narrow range of the energy spectrum is also inhibited by chemical, electrical and radiant imbalances that alter the life cycle of the cell. A cell's life cycle ends when natural and induced imbalances take place and the cell is no longer able to transfer energy within the narrow range required for their replication. An analogy of the life cycle of a cell might be the acid battery: thin lead cells produce sky rocket results in chemical energy transfer to electricity in the short term versus thick lead cells that have a long term ability to transfer chemical energy into electricity. In both batteries, there are no assumed induced imbalances. In the electrolyte world of the biological cell, it is the gene pool that determines the cell's ultimate life cycle. This gives meaning to the phrase, 'Get the lead out', to engage in a sky rocket life style.

The strata within each biological cell are on unique spiral path segments that are influenced by energy spectrum transfers. At birth the strata are in a very tight high energy state around their axis of direction dominated by gravity. The energy spectrum transfers to the cell strata include gravity and any excess gravity received is passed to their respective nucleus that provides the dominant axis of direction for the cell strata. At each replication of the cell, the cohorts in the strata loosen their tight spiral paths as events along their parallel time lines start, in miniscule amounts, absorbing some of the energy spectrum instead of transferring it intact to their nucleus. Maturity is where the weakened energy spectrum exchanges, particularly gravity, are materially reduced within the cell to their nucleus. This causes a reduced frequency of replication of cells and the gradual atrophy of all specialized cells in the corporate body.

An example of cell atrophy is in the specialized connecting tissue to muscle cells. With the inability of muscles to contract, they become atrophied and the body cools as the heat transfer from muscles slows down. To maintain the energy exchange spectrum, the body needs an increasingly induced warm environment. This loss of the energy spectrum, which includes gravity, precipitates age related weight loss. The death of the corporate body marks the end of the life cycle of all of its cells, and the remnants of the energy spectrum are released from the corpse as a transfer to its environment.

The DNA double helix mechanically tracks changes to the cells based on the history of replication. The human genome consists of millions of inactive segments that continue to replicate. This is the mechanical tracking of our historical genetic drift. They are not applicable to our current cell structures. The replication in the DNA double helix is also part of the energy spectrum transfer when a cell reaches its energy threshold and sheds its excess gravity along with a share of its strata that include DNA.

Underlying the relative terms of the laws of physics are the six absolutes of TUT. For this reason it is necessary in the discussion of biology to make this linkage to TUT with the laws of physics.

The laws of physics apply to the cells in our human body and all existing forms of life. These cells have either adapted to these laws as their environment changed, or perished. The cells that have adapted also create niches in structures that obey the same laws, and they become mutually bound together. The laws apply universally to the respective corporate bodies the cells occupy. The corporate bodies, the body of unified individual biological cells, of mammals, birds, other terrestrial and marine creatures benefit from this cellular evolution and by individually adapting to the laws of physics in their particular environment.

The external influence of bacteria and viruses overwhelm a cell's inherent inclination to adapt and in turn the cell's ability to obey the laws of physics. This domination ultimately inhibits the cell from functioning corporately.

The human corporate body's evolution has made it adaptable to many environments. But the corporate body is at peril when incompatibility is induced in the body's cells preventing them from obeying the laws of physics. This inducement can be personal, cultural or political. Personal and cultural inducements such as smoking or excessive use of recreational drugs may not be readily apparent to the corporate body, but the cells lose their ability to meet the laws of physics that sustain them. Political inducements are the massive environmental changes that result from government decisions, or indecisions, that inhibit the body cells to function according to the laws of physics. Liberal policies rely on a relatively short term mantra, 'Dilution, of hazardous material, is the solution'. Conservative policies are defended by their mantra, 'It's the economy stupid!'. In either case the political policies undermine the sustainability of human life on the planet earth.

In fairness, as an auditor, I have encountered one area of government policy that is guided by the laws of physics. The area is public safety, such as construction, transportation, manufacturing, and health and child care services. My philosophical perspective is that all personal, cultural and political inducements must be subject to the laws of physics. Why? Because it is something we have control over, our cells are totally reliant on their respective corporate bodies being subject to the laws of physics.

Strides

Newborn ants sail aloft in the soft summer sky
Until a loss of wings end their airborne rides.
Now confined to the earth to ever more crawl
Where each inch forward takes many strides.

The birds can soar in the great wide open skies
Across earth's continents and any expansive ocean
Compared to the many strides of creatures below
Who travel on four legs in perpetual motion.

The extension of man's physical stride on earth
Allow for planetary travel to our moon so fair.
He reduced the time to travel around the globe
From every place on earth to almost anywhere.

But many strides are needed to grasp our position
Along the spectrum of an infinite universal space.
With our universe being so large and yet so small
The centre must be occupied by the human race.

—*Opa's Journey…*

Topography

The topography of our solar system is well researched and mapped. The solar system's position in the dominant spiral path of our galaxy has also been researched and its probable position mapped. But to focus on our galaxy is like focusing on a grain of sand rather than observing the sand dunes that spiral out into absolute space. Driving on a highway in our relative world the focus of our roadmap is on the asphalt road, not the aggregates in the asphalt. However, the fodder for most modern science is in the minutiae. If the minutiae are the fodder of science, then the minutia is the fodder of TUT. In the relative world, the sky is up and the earth is down; there are two sides to a coin, and only one side to a Möbius strip. A point of existence has no up, down, nor sides. Therefore, a point of existence has no topography but its cohorts provide the minutiae in the science of topography.

The earth's spiral path around the sun is variable. According to TUT propositions, an increase in energy transfer tightens the spiral path and a decrease in transfers loosens the spiral path. As the earth approaches its perihelion, energy transfers from the sun which exceed the earth's transfer to the sun, are absorbed by earth. This excess energy is released to the sun as the earth moves back again to its aphelion position. Switching on and off is a TUT proposition. The point of switching is where the sun's energy transfer is equally countered by energy transfer from the earth. The switching of energy transfers at the points of the earth's perihelion and aphelion provide the topography of the earth's variable spiral path relative to the sun.

Binary stars have mutually close variable spiral paths and energy transfers between them are extreme because of their proximity to each other that, according to TUT propositions, allows large frequencies of constant tiny tight gravity transfer encounters. Because the energy spectrum's axis of direction is gravity, there is a very large release of energy from the binary stars as their individual axis constantly changes along a new mutual axis of direction. According to TUT propositions, these dominant changes in the axis radiate the combined energy spectrum of each star outwards, including their large magnetic fields.

TUT propositions also state that each point of existence is unique in a spiral path and its position cannot be occupied by another point of existence. For these reasons, the spiral paths of both stars and planets are also unique. In absolute space, stars and planets collide with smaller pieces of debris that lack enough internal energy transfers on their spiral paths to maintain unique orbits. These collisions result in a dramatic release of energy because, according to the propositions in TUT, the spiral paths of the debris are abruptly changed by the event. The debris cannot respectively occupy the same steady state points of existence of the stars, planets and their moons. The impacts of the largest of this debris on the land masses of the earth and its moon have observably changed their topographies in the relative world.

Figure 24

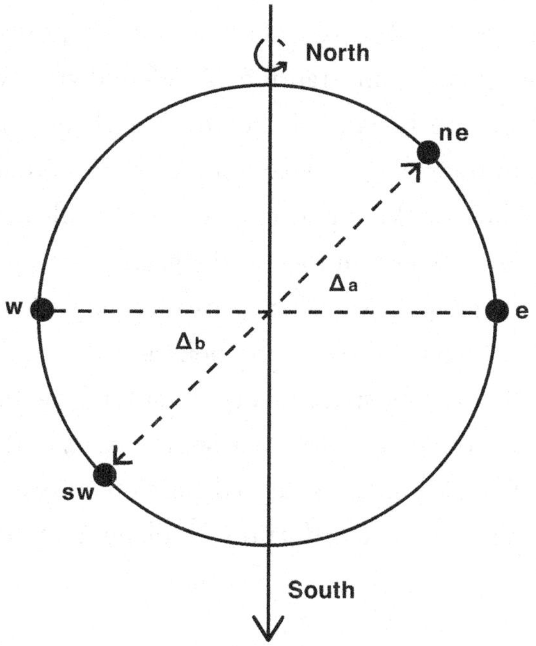

The spin of celestial bodies, according to TUT propositions, is where the energy transfers from the core create larger and larger spiral paths until they reach the surface. The core of planet earth has a relatively tight spiral path and energy transfers to the magma and the surface must follow their respective loosening spiral paths along their mutual dominant axis of direction from the core. In TUT, it is the stratified cohorts of existence that move individually on their respective unique spiral paths through space. This is observable in the relative world as spin as each stratum in a celestial body follows their respective spiral paths along their mutual dominant axis of direction.

On the bisected celestial body in *Figure 24*, the time line of an energy spectrum transfer of radiated gravity, where angle $\Delta_a = \Delta_b$, observable vortices are created at **ne** and **sw**. Where angle Δ_a and Δ_b = zero at the equator, **w** (west) and **e** (east), there are no observable vortices. In philosophical terms, the centre of the bulge on the surface of the equator is like the top of Mount Parnassus: We can go down only one side at a time but not both sides at the same time, or we can simply remain on the top. As long as we stay on the pinnacle at the top of Parnassus, our position is stable. Moving to either side our position becomes unstable as energy spectrum transfers of radiated gravity increase.

The bulge on the surface of the equator absorbs the radiating energy spectrum transfer of gravity from the core so that there is less gravity on the line of the equator. The bulge is also shaped like a converging meniscus lens that focuses the gravity transfers to absolute space over the principal axis of the lens that aligns with the line of the equator. The focus of the lens sharpens as it reaches out from the equator line into absolute space in a plane around the entire celestial body. The locus of the focal points in absolute space, or the pinnacle in the Parnassus effect, is fundamental to the architecture of the universe.

The observable Parnassus effect on the planet earth is the stability in the trajectory of GPS satellite orbits. The orbits are not pure ellipses over the equator. The planet Saturn, which has the largest bulge in our solar system at its equator, has enormous stable rings of trapped space debris on the plane of its equator.

The stability of the Parnassus effect applies universally and is observed in the shape of galaxies. Their arms extend, from the equator of their respective cores, outward in a plane along a path of stability. This suggests that for practical reasons inter-galaxy travel would require following this stable plane to facilitate a journey to another galaxy. According to TUT, in *Figure 24* the angle of travel on the plane of a galaxy's equator, like the equator of our Milky Way, must approach zero.

The topography of our observable universe, which TUT defines as being relative, is ultimately changed as our universe expands on its immense spiral path. According to TUT propositions, universal expansion is based on movement from existence to non-existence within the spiral paths of space. As celestial bodies move outward, they are releasing energy to points of non-existence. Absolute space is infinite and where there are fewer points of existence the spiral path is constantly variable.

The whole spectrum of energy moves to fill theses points of non-existence creating more variable spiral paths. Remember TUT sees time as being fixed. Time lines are both infinite and pass each other on infinite angles. In infinite absolute space and time there is no randomness of events. The very term infinite means we are defining all of absolute space intersects with time. Therefore, infinite includes all events on time lines not just segments of time lines. Intersects by spiral segments on time line segments are statistically random events but are not random in infinity. This concept is important to understand changes to the topography of our known and relative universe where events are random.

Figure 25

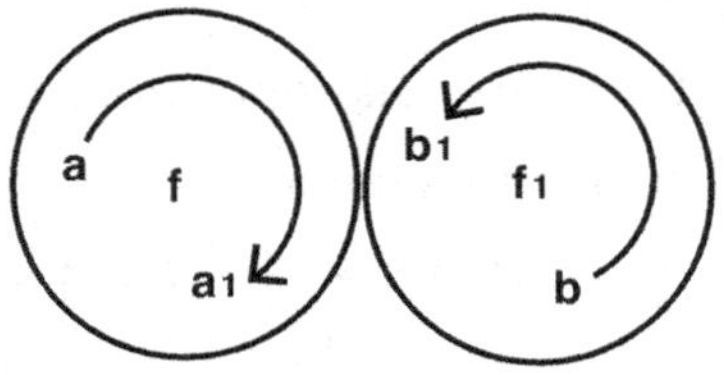

Axis of direction for the direct mechanical transmission of energy on gears

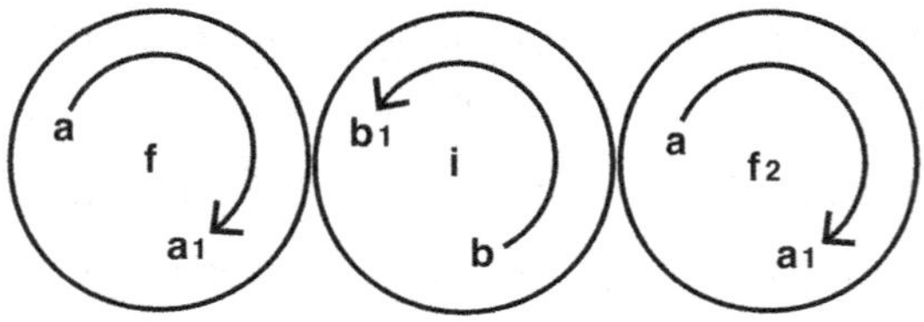

Axis of direction for the mechanical transmission of energy via an idler gear

To audit business systems and processes, the auditors prepare flowcharts or document what is taking place within a short but current time frame. Within the dynamic business environment this creates snapshots of the organization, its systems and processes that are currently in place and allowing an objective audit opinion that is based on observed current practices not on speculative past or future events. However, the audit opinion also includes events that could have potential future risks to the business. Applying this audit approach to charting and documenting the topography of our universe, I'm able through TUT to reveal the dynamics of events on parallel time line intersects where the observer is always at the centre of past, current and future events no matter where the observer is located on the universal spiral path. Since we have vision only in one direction, the topography of our universe would require virtual snapshots of what is past, present and future from our centre. According to TUT, the past, present and future is challenging but determinable.

The magnitude of a universe's expanding spiral path is so great that it approaches the point of being a linear path. At the point of linearity, the inherent energy spectrum is released to points of non-existence on spiral segment paths as events on parallel time lines with angles not shared by this universe. In our relative world, this climax of events is observed by us as a black hole.

Where the energy spectrum reaches its state of equilibrium, the expansion ceases and cohorts of existence coalesce as events along these new parallel time lines creating hydrogen clouds, with gravity as the axis of direction for the ultimate topography of a new universe. The new universe will have symmetry and topography with the chaotic history of our existing universe because space intersects with time over infinity are not random. TUT provides the means to reasonably determine the relative topographies of events then, now and to come.

The gearing of the direct mechanical transmission of force in *Figure 25* from its energy source on gear **f** has its axis of direction **a** to **a1** reversed on gear **f1**. The new axis of direction on gear **f1** is **b** to **b1**. To eliminate this reversal, an idler gear **i** is used to transfer the force and maintain the axis of direction from **f** to **f2**. The high speed mass less and charge less neutrino particle is analogous to an idler gear that does not have its own intrinsic load. The neutrino only serves to transfer the energy spectrum along with its inherent axis of direction. The idler gear can also be used to adapt a large slow moving gear to a small fast moving gear, or vice-versa. The neutrino transfers larger spiral paths to tight spiral paths or vice-versa and maintains their axis of direction. Neutrino cohorts efficiently facilitate the energy spectrum transfers to points of non-existence on spiral path segments.

There are an abundance of neutrino particles in absolute space and all have close linearity with their parallel time lines. Their points of existence are on extremely tight spiral paths, and the cohorts of each particle blink on and off as unique events on their parallel time lines as a result of their catalytic role in energy spectrum transfers.

The parallel time lines for each particle are like parallel train tracks. The high speed particles require close miniscule angles to make very subtle energy spectrum transfers between points of non-existence on spiral segment paths. Using the train track analogy requires a very small angle for high speed trains to switch tracks. In the same relative terms, high speed neutrinos would also require great distances for these particles to change their track where the subtle angles and positions of other parallel time lines and spiral segment paths are actually in place for these events. This allows the neutrinos to penetrate celestial bodies with low probabilities of energy exchange due to the high level of points of existence in these bodies. Energy exchange events are more frequent if there is a high level of non-existence that is not normally found in celestial bodies. To physically make an observation of a neutrino event on earth, would require a very large body of liquid such as water and a lot of patience. However, where these particles meet, the spiral path of a universe at its climactic point of linearity with parallel time lines, they enter a very large oversized virtual train switching yard and readily transfer the energy spectrum cohorts of the new universe being created.

Grains of Sand

The grains of sand on the desert floor
Give the appearance of a solid place
When the grains move imperceptibly
And other grains shift into each space.

Shaped by the ceaseless winds or seas
They progressively find a sheltered spot
Covering everything meeting their path
In dunes under the blazing sun so hot.

Each granule's distinct coarse exterior
Hides a hard crystalline core that is built
To assure their long existence on Earth
Before they grind each other into silt.

Earth's creation from a cosmic dust cloud
With its crust that is pummeled into sand
Recreates more clouds of the cosmic dust
From whence our Universe first began.

—*Opa's Journey...*

Conclusion

Points of existence and non-existence are metaphorically yang and yin, or male and female. The solid cohorts in the male points of existence would be inert without the passive female points of non-existence that allow energy conversion to take place. Energy conversion is the very source of our creation that includes our creative and critical thinking that gives us art, music, dance, literature and science. Non-existence is pervasive throughout our universe and dominates the atoms, molecules and cells in our bodies.

Having separated the untruths to arrive at the truth, I can finally say to Sherlock Holmes, 'Elementary my dear Holmes, TUT, philosophically and metaphysically, reveals that our pervasive universal God–is female. She is everywhere and in everything. Primitive societies worshipped the passive moon, stars, planets and earth. They were absolutely right. She is there and also in all of us. She is an ethereal spirit whose mere breath changes the universe by annihilation to create space for all that is new'.

Furthermore, if I could, I would assure my mentor Lawrence Sawyer that TUT provides the means for an audit of the universe, both physically and metaphysically. And that he and Archimedes have inspired me to say with certainty, 'Give me a small enough surgical blade with an edge of single points of existence, and a place to hold it, and I will dissect any matter in the universe'.

One of TUT's conclusions is that the tools of the scientific method in our relative world primarily rely on inference for calibration rather than absolutes. Inference is a useful tool in our relative world but is a blunt instrument to apply universally. For example, one of these most commonly used tools is a thermometer that is calibrated to indicate the boiling point of water to be 212 degrees Fahrenheit. But this is only at mean sea levels, not at other elevations. With our planet's current rising seas a recalibration would be required each century in the future.

And it would still not apply to other planets or their moons. In the calibration of energy, or time and space, inference applies loosely rather than absolutely. Since not all events are known, it adds an element of 'maybe'. Boiling, according to TUT, is no more than an energy spectrum transfer that is universal and absolute rather than something based on the inferences found in the traditional scientific method.

TUT defines the universe as virtual reality. That means what occupies the universe are also virtual realities that shift between existence and non-existence. This is why all that is possible in the universe is possible! The only limitation is in our ability to imagine what is possible.

Finally, it is important for us to understand, acknowledge and protect our web of life. Our body cells are fragile and we, the corporate bodies, have a respective symbiotic relationship with each one of them. If abused by recreational drugs, lifestyle, hostile environments and induced stress, the specialized cells will respond by closing down our organs and life ceases. The conclusion is that we have little control over our relative world. The TUT propositions indicate we are definitely not in charge of our points of existence within us or the rest of the universe that surrounds us. Infinity has no limits but our universe does and we are bound by its past, present and future fate. Our universe is on an immense variable spiral path, according to TUT, and therefore the only constant we can expect to encounter with certainty in the relative world we have created on our planet is change! This change will ultimately lead to the annihilation of our universe and the rebirth of another universe. The positive news from TUT is that we are certain to be there once the chaos of the new universe subsides.

Notwithstanding the rigorous absolutes of TUT, we should cherish our very short presence in a relative world and learn to live with its vagaries. We need to embrace the beauty of nature, the visual arts, literature, music and the philosophy of science. We exist as their cohorts. So let sunrises bring joy to our lives. And let sunsets bring a special awe.

A Sailor's Delight

A glorious sunset over the sea
Sends a ripple of awe to all
Who sail or must stay ashore,
Silhouetting those ships at sea
In such crimson splendor
That transcends water and sky,
As the sun journeys into the sea
The red skies we see at night
Are not only a sailor's delight.

—*Opa's Journey...*

Questions and Answers ...
According to TUT

When I was six years old and showing off my grade one school marks, my fifteen-year-old, brother Ken said to me, 'If you're so smart then answer these three questions. How high is up? How low is down? Where does a flame go when it goes out'? His questions humbled me for almost seventy years until I fully developed The Universal Theorem (TUT). According to TUT, there is no up or down. There are only events on time lines that cross, but do not intersect and extend to infinity. For example, in our relative world, we might flip a coin and choose heads up or tails down. According to TUT, if we took this coin's points of existence on the top surface, the cohorts would have no dimensions because their points of existence flash between existence and non-existence. To have a choice of heads up or tails down, requires three dimensions that only exist in our relative world. The multitude of TUT events on either face of a coin are seen by us as one single event and is the basis for our relative perception of reality. The philosophical question becomes why do you have to see something to determine if it exists? The short answer, according to TUT, is you don't!

If we apply TUT to a flame, its energy spectrum transfers radiate outwards in all directions to infinity unless intercepted by something. For example, if the flame from a candle is intercepted by our eyes, both the flame and candle are retained by our brain as a virtual image. This image remains long after the flame is extinguished. Philosophically, for as long as we live, the flame never goes out! Just think or say *candle,* and the combined image of the flame and candle is retrieved by our minds.

My conclusion is that it was my brother Ken's non-critical response to his show-off baby brother that instilled a boyish intellectual curiosity that has lasted me a lifetime. Thanks Ken!

If TUT is truly universal, then it should be able to answer any reasonable question prefaced with the word, why. The following are random questions about a few common, yet disparate subjects, to illustrate this point.

Question

Why does my heavy fry pan appear to get cold before it gets hot when I first put it on the stove?

Answer

According to TUT, energy spectrum transfers to a spiral path segment cause its path to tighten around its axis of direction. The energy transfers radiate back and forth from the spiral segments' axis of direction. When placing a cast iron pan at room temperature on a hot stove, the axis of direction of the spiral path segments on the bottom of the pan accepts the energy transfer as the spiral path tightens. Since this transfer is also upward to the interior of the pan, each spiral path segment above begins to tighten around its axis of direction that in turn draws heat transfers from below and above. Once these transfers reach the surface of the pan's interior, there is a differential in energy transfers between the bottom and interior surfaces of the pan. This is why initially the interior surface of the pan drops below room temperature as the spiral path segments close to the upper interior surface start tightening around their axis of direction and draw energy transfers from the interior surface. An energy transfer differential is always maintained because the axis of direction radiates energy transfers in all directions. Once the energy transfers reach, and are maintained, at a cooking temperature, the bacon and eggs can be added for breakfast.

Question

Why do we have high and low atmospheric pressures in weather systems?

Answer

When I was in boot camp, while serving in the Royal Canadian Air Force (RCAF), 411 Fighter Squadron, we studied cloud formations and centres of high and low atmospheric pressure.

In simple terms, we learned that placing our backs into the wind from a high pressure system, the centre of the system was to our right side. With the wind on our backs from a low pressure system, the centre of the system was to our left side. In TUT all motion follows a spiral path.

In a high pressure system, gravity transfer radiates out from the earth on tiny spiral paths to the atmosphere. The direct absorption of gravity transfers by the atmosphere creates the high pressure. The evidence is in the wind spiraling down from the gravity transfers that create the high pressure.

The sun's energy spectrum transfer to water surfaces results in evaporation, humidity and cloud creation. The energy transfers include gravity in the water molecules. As the magnitude of the clouds increases, their gravity transfers are focused on the concave bottom of clouds back to earth. When these focused gravity transfers from the clouds exceed the radiated gravity transfers from the earth's convex surface, the air spirals upwards creating a centre of low pressure. The spiral path tightens while absorbing energy exchanges or expands when releasing energy. So in high pressure weather systems, the earth's gravity is absorbed by the atmosphere, and in low pressure systems the gravity in the clouds is returned to earth as wind, including updrafts and precipitation. High pressure weather systems tend to be large stable spiral paths while low pressure weather systems tend to be small unstable spiral paths. The wind is also independently affected by large bodies of water, particularly at their shorelines. The differential sea and land temperatures can create onshore or offshore winds. This is why our Florida beach house, on a hot day, is cooler from the onshore winds than houses situated only a short distance inland.

Question

Why is ice so slippery to walk on?

Answer

When I moved to Sault Ste. Marie, Ontario, many years ago I decided to go ice fishing on the glacier blue ice of Lake Superior. The morning temperature was minus twenty degrees Fahrenheit, and I had absolute firm traction walking on the ice. I took my axe and struck the ice several times and was only able to dislodge ice the size of toothpicks. The ice had traction and the consistent hardness of granite.

TUT proposes that an event on a time line can only be occupied by a single point of existence. Water vapour releases energy as it cools and condenses into water then further condenses into ice. During these energy spectrum exchanges, the points of non-existence become more occupied. Ice has the least points of non-existence and these remaining points of non-existence are further reduced as the nucleus of each atom in the ice transfers its remaining energy spectrum. TUT also states that points of non-existence allow universal movement. This is why the ice on Lake Superior was like granite. The leading edge of my axe could not reach enough points of non-existence to make a hole because most of the points of non-existence were occupied by points of existence.

Therefore, the answer is that ice is no more slippery than granite! But put water, which contains lots of points of non-existence, on the ice and the water film between your boots and the ice will become traction free and very slippery. This is also true for ice skating. A film of water on the ice provides us with more effortless skating. Freezing rain on a ski hill will also provide us with the downhill hair-raising thrill of the proverbial, 'slippery slope'.

Question

Why doesn't there appear to be any certainty in the world?

Answer

The poem *Pure Perfection* at the end of the section The Philosophical Context lists our world's imperfections but arrives at the conclusion that '*... the sum of all imperfection... is pure perfection!*'.

In quantum mechanics, the imperfections found in Newton's laws of motion are given an uncertainty factor. In TUT the infinite angles on time lines that cross each other without intersecting create random events in our universe that give rise to uncertainty. Also, according to TUT, Big Bangs instantly create order in each new universe. The disorder begins to unfold, as the new universes cool and coalesce on their respective spiral paths. Our points of existence have travelled these paths over and over again. It is from this ultimate sea of disorder that we acquire the necessary cohorts to become what we are. Infinity, by TUT definition, includes all universes and possible events, and therefore there is no randomness It appears that the sum of uncertainty through infinity is certainty. However, at this eventful moment of our human existence, we are confined to our universe with its growing uncertainty–being a certainty!

QUESTION

Why can't we physically go back in time?

ANSWER

Ancient civilizations used ruins of more ancient cities as building material. Stone foundations and walls were used on site to build new cities on top of the older cities. Archeologists give us a sense of going back in time by retrieving artifacts from digging down through different layers of civilization. The living inhabitants of these ancient cities no longer exist, but their ancestors, you and I, are alive and well.

According to TUT, we leave a biological trail of debris as long as we live. From the moment of conception, our mother's blood feeds our cells to sustain our growth while removing the debris from cells that our growing fetus discards. At birth we begin discarding our own debris of feces, urine, sweat, blood, breath and dead skin tissue. Once discarded, the debris begins to lose its points of existence acquired as cohorts by its human hosts. The remnants of these cohorts, like the building blocks of ancient cities, become the building blocks for new biological structures. The restructuring events take place on trillions of parallel time lines from an equal number of unique intersects by spiral path segments. This follows the TUT proposition that no two points of existence occupy the same point of intersection. Therefore, if all of our points of existence are occupied in our cellular structures, then any debris they discard allows spiral path intersects on the time lines to form new points of non-existence to maintain the energy spectrum exchanges in our cellular structures necessary for cell replication.

For this reason, our body's biological time clock is forced to be in the present. We can remember the past, plan for the future, but only experience our physical presence by being here. Remember when we were children, demanding our busy mother's attention and she quoted that old adage, 'I can't be in two places at once!'. It took over fifty years, for me to develop TUT to ultimately prove that our mothers were absolutely right! According to TUT, and our mothers, we can only physically occupy the present.

Question

Why is there no perpetual motion?

Answer

When I was a child, I received a gift as evidence that perpetual motion was possible. It was a vacuum bulb with delicate silica wafers that revolved around an equally delicate shaft, all of which was mounted on a sturdy base. Streams of photons striking the wafers caused them to spin. Because there are an abundance of photons darting about, it appeared to never stop rotating. But it was not by definition perpetual motion as the speed varied with the intensity of light and, therefore, stopped when I mischievously moved it to a dark corner of my closet.

TUT's universe is in a perpetual state of motion because of energy spectrum transfers along spiral paths that extend to eternity. As long as we live, we will experience the apparent perpetual motions of the stars, planets and moons. But this experience is based on our relative world experience. It is only the points of existence in the universe moving to points of non-existence on spiral paths that, by definition, are in perpetual motion. Spiral path segments and their cohorts that appear to us to be stable single events in the relevant world, however, are influenced by the dominating variable spiral path of the universe that is the source of perpetual motion. Also, according to TUT, this perpetual motion of points of existence in spiral path segments cannot be observed. The single intersect of a spiral path segment on a time line segment is unique, as it has no dimensions, no mass and thus is unobservable. Therefore, there is no directly observable perpetual motion.

The conclusion is that perpetual motion events are the infinite universal and invisible *means.* But in our relevant world, the *ends* we observe are merely parts of spiral path segments that restrict our observable *ends* to multiple events appearing to us as single events.

Question

Why does one plus one equal two?

Answer

There is a caveat that to equal two each of the ones must also be equal to each other. Using an algebraic example, $1a + 1a = 2a$, the same caveat applies. By substituting a with *apple,* the apparent solution is 2 *apples*. But what if one apple is red r and the other is green g? If each apple is not equal to the other, the answer of two equal apples is incorrect. What we really have is $1ra + 1ga = 1ra + 1ga$. This would apply to any substitution for a.

A circle is defined as having any point on its circumference at an equal distance from the centre of the circle. TUT proposes distance is curved and the dominant spiral path of our universe is variable. That means while drawing a circle, the curve becomes distorted and the points on its circumference would not be equal distance from its centre. Therefore, all geometric figures such as triangles, rectangles and conic sections do not exist because the distance of the lines forming their shapes are also curved and the figures are distorted. The distortion is not visible when the figures are drawn on a piece of paper, but if their shapes were extended into absolute space, the distortion would be immense.

Philosophically, it is also interesting to note that equality is imbedded in our Western cultures that rigorously pursue free enterprise that thrives on inequality. The problem is the application of the word *equal*. Even a gallon of gasoline in America is not equal to another gallon of gasoline, unless they are both at the same temperature, elevation and the ethanol in them has not absorbed water from the atmosphere.

The conclusion is that, based on TUT, the definition and symbol of *equal* only applies universally to the deemed angles where fixed time line segments cross; therefore, its application in the relative world should be used with caution.

Question

Why is alcohol sometimes called absolute?

Answer

The word absolute, by definition, means free from imperfection or having no restrictions, exceptions or qualifications. *Webster's* dictionary provides exceptions for the use of this word including absolute alcohol and absolute space.

Absolute alcohol is the common name for ethanol, or grain alcohol, that is at least 99% alcohol by weight. Absolute alcohol is anhydrous and readily attracts water vapour if exposed to the atmosphere. Water in alcohol creates unwanted reactions in certain chemistry applications so it is important it has the least amount of water. This places a limitation using the word absolute when the ethanol is only 99% alcohol by weight.

Absolute space, on the other hand, has a relative meaning which, coincidently, applies to TUT. Webster dictionary defines absolute space as physical space independent of what occupies it. In TUT, space and its propositions are one of the six absolutes that are universal.

Absolute alcohol is not universal because it based on the maxima of at least 99% ethanol by weight. The maxima are conditions and, according to TUT, the word weight, defined in limited relative terms on earth, is not a universal measurement. The TUT absolutes have no conditions, or limitations, and apply universally.

Question

Why do meteors glow when entering the earth's atmosphere?

Answer

Meteors are debris originating from random collisions of fragments of celestial bodies not having stable, fixed orbital paths. The collisions transfer the energy of their bodies to the debris; this radically changes their respective orbits into a multitude of individual orbits. This often places the debris in the paths of stars, planets and moons that have stable, fixed orbital spiral paths. The debris in the path of earth has been travelling on long journeys through absolute space. TUT defines absolute space as having an abundance of points of non-existence, and the debris' spiral paths are large and stable.

In a collision with earth, even a glancing collision, the debris' orbital path is abruptly changed into variable spiral path segments by the atmosphere, thereby releasing spectacular energy spectrum exchanges for observers on earth, day or night. Since the energy exchanges are radiated outward from the axis of direction in the core of the debris, the core may remain intact until the orbit stabilizes. Debris consisting of compounds disintegrates creating what are called meteorites. Metallic elements in debris contain large numbers of points of existence that are sacrificed to the atmosphere at the surface of the debris allowing the core to reach the crust of the earth. Once imbedded in the earth's crust, it becomes an integral part of earth's stable, fixed orbital spiral path.

In conclusion, these spectacular graduated streaks of energy by meteors and meteorites are wonderful visual examples of universal energy spectrum transfers with the earth's atmosphere. Where there is no atmosphere on other celestial bodies, there are no glowing streaks from meteors. The meteors become deeply imbedded in the surface of these respective stars, planets and moons by their instant energy exchange upon impact.

Question

Why can helicopters move in different directions?

Answer

Where a helicopter is parked in a stationary position on the ground, TUT proposes that the cohorts of the points of existence of the helicopter are in the steady state of our planet and thus shares its axis of direction. The movement of the helicopter is a result of energy spectrum exchanges. This is accomplished by the spiral path created by the main rotor blades. Changing the pitch of the blades tightens or expands the spiral path, and the helicopter creates its new dominant axis of direction. The tail rotor creates its own spiral path that provides sufficient opposing spiral energy transfers to the hull of the helicopter to counter the spiral energy spectrum transfers on the hull created by the engines. However, the spiral path of the tail rotor is subject to the dominant spiral path of the main rotor blades. As the dominant spiral path of the main rotor tightens, or expands, the helicopter vertically rises or descends respectively. The helicopter changes its axis of direction according to the tilt given to the main rotors thereby moving the helicopter from vertical to lateral directions.

The atmosphere is dominated by gravity, and it is the change in the spiral path of the surrounding air created by the main rotors that is the source of the energy spectrum exchange. For a helicopter gravity is the dominant energy spectrum exchange. For this reason, the helicopter could be considered analogous to an anti-gravity machine. This applies to aircraft, ships, boats and rockets that derive energy exchanges from the tight dominant spiral paths created by their respective propellers or propellants.

A helicopter's gradient movement in different directions between vertical and lateral apply in earth's atmosphere but not in our low gravity stratosphere nor in a low gravity atmosphere on other celestial bodies.

Question

Why do time lines in TUT cross each other at any angle but do not intersect?

Answer

Let's draw any straight line on a piece of paper where the line represents a time line segment. Now draw another straight line at any angle across the first. We have just demonstrated an application of TUT in relative terms. The second line crosses over the top of the first line at any angle but does not intersect it. If we drew the second line with no angle relative to the first line, then it would be parallel to the first line and, by definition, would not intersect the first line. This includes drawing the second straight line directly on top of the first straight line. These lines do not intersect and therefore, by definition, are also parallel.

To answer the question–Why these straight lines representing time line segments do not intersect? –it is because they were drawn at different times where each line represents a unique fixed time line segment. Since TUT propositions are universally absolute, we are bound by them. No matter which straight line we draw, the moment of placing the pen on the paper is unique. To finish the first line, to paraphrase an old adage, 'the pen must move on', thereby creating a multitude of unique events until the pen is lifted from contact with the paper, leaving a straight line representing a time line segment. To draw the second straight line, the pen moves on once again but creates a new multitude of unique events not shared by the first straight line.

Question

Why is electrical current delivered to our homes as 60 cycle alternating current?

Answer

When I was a young boy my father, who built radios for a living when he was a young man, was excited about the switch from 40 cycle to 60 cycle alternating electrical current at our house. And he pointed out to me the visual change when we turned on our lights: At 40 cycles there was a physical sense of the cycling taking place as the light wavered. At 60 cycles the light bulbs provided a bright continuous glow.

This was a relative, and relevant, demonstration of TUT absolutes. The energy spectrum transfers at 40 cycles were akin to sensing multiple single events flashing in and out of existence. At 60 cycles we observed only one event. So when we make an observation from a natural light source, or reflected light, we are actually witnessing an infinitely large number of TUT events as a single event. And that is a good thing! Do we really want to read our favourite book under strobe lights? It is an application of TUT whereby we are not necessarily seeing what we are getting thanks to our 60 cycle alternating electrical current. At 40 cycles we sense the minutia of energy spectrum transfers and at 60 cycles we sense the minutiae as a single event.

Question

Why does space travel debilitate the health of astronauts?

Answer

When Canadian astronaut Chris Hadfield returned to earth on May 13, 2013, after 146 days orbiting earth, he was 'tottering around like an old man', according to an article in *The Globe and Mail* (May 16, 2013). The symptom relates to how our human bodies function best within a narrow range of gravity on the surface of the planet and is necessary for inducing the biological replication of our cells. Long periods of low gravity, or lack of gravity, slow the energy spectrum transfers within and between these cells. The decline of these energy spectrum transfers result in an accelerated atrophy of all cells in our body.

My conclusion, based on TUT, is that all internal cohorts of existence respectively lose their ability to transfer energy when an astronaut leaves earth's narrow range of gravity, during space travel, and their respective body cells cease to replicate. Our body cells also lose the ability to replicate as we reach old age. This explains Chris Hadfield's symptoms of old age upon his return to earth, and the reason why I started tottering around at the age of 75 years old!

Author's Note: For more recent TUT questions and answers go to the Q&A page on my website, www.davidtovell.com. There is also a link for you to provide your personal comments on the Q&A topics.

Acknowledgements

I would like to recognize and offer my gratitude to the following people: my editor, Laura Abbott, for her patience and editorial skills; Stephen Waicus, who took many reiterations of my very rough sketches to produce understandable and visually accurate digital figures, and Cindy Cake, for her professional graphic designs. And, finally, to Beth Crane, for her outstanding text design and layout for our book. Their teamwork, dedication and skills were very much admired and appreciated.

Finally, I would like to thank the Astronomy & Astrophysics Library for lending me a key to access their collection of journals and books in the McLennan Physical Labs Building at the University of Toronto many years ago. Over a period of ten years, one day each week, I would smuggle my lunch bag into the library for my lunch and learn, 'happy hour', of inspiration while perusing their current published material.

About the Author

David Tovell was a professional auditor for forty-three years, whose clients were ranged from small to large businesses in Canada and the United States. During his career, he wrote several articles for major trade publications, including *The Internal Auditor Magazine, USA, and CGA Magazine, Canada*. After retirement, he published a collection of his poems in *Opa's Journey*… David and his wife Margaret live in Toronto, and spend their winters in Florida.

The following poem was written by the author to portray, in TUT absolutes, the virtual reality of his own existence beyond his careers as an accountant, auditor and writer.

I Am...

I am now and then a point of existence in space and time
With no dimensions and in a state of an inert condition
Until blinking out of existence between parallel time lines
And curves of space that create my continuous transition.

I am joined by cohorts readily sharing my unique place
By filling trails of my non-existence as I in turn fill theirs
Into a steady state of co-existence to create the shape
Of events along our mutual time lines in parallel layers.

I am co-opted by numerous cohorts on my axis of direction
Transferred by a shape of space between what we require
Within strata to maintain separate points of existence
For the adaptive cohorts of human attributes we acquire.

I am bound by the constant change in the shape of space
Ultimately transferring my cohorts that made me human
Until my point of existence finds a place in a new universe
So that I can explain once again, in a poem, why I am...

Made in the USA
Monee, IL
04 November 2026

67378671R00056